MEDITERRANEAN DIET COOKBOOK FOR BEGINNERS:

The Complete Mediterranean Meal Prep Guide for Weight Loss with Delicious and Easy to Prepare Recipes in a Balanced 30-Day Meal Plan

Introduction 3

Chapter 1. What is the Mediterranean diet? 4

Chapter 2. History and benefits of Mediterranean food 5

Chapter 3. How To Lose Weight By Eating Healthy? 13

Chapter 4. First two days of detoxification from junk food 15

Chapter 5. 4-week meal plan 18

Chapter 6. Breakfast 23

Chapter 7. Lunch 42

Chapter 8. Dinner 64

Chapter 9. Dietary desserts 81

Chapter 10. Desserts for special events 100

Chapter 11. Daily snacks 122

Chapter 12. Eating out 145

Chapter 13. Recipes for special events 147

Chapter 14. Bonus: Recipes for Air Fryer 163

Chapter 15. Bonus: Traditional Italian recipes 183

Conclusion 212

Introduction

As the name says, the diet belongs to the countries surrounding the Mediterranean Sea, including Greece, Italy, Spain, and France. And after studying the short history of this part of the world, we do realize the vast cultural richness of these states and the diversity of the people and their different lifestyles. One common thing which connects them all is the food they consume and the way it is consumed. Legumes, vegetable, nuts, fish, grains, cereals, fruits, beans, and good fats, these are things which mainly forms the Mediterranean diet. So, it guarantees a good amount of fibers along with all the macronutrients like carbohydrates, proteins, fat, etc.

Several scientific studies have proven that the Mediterranean diet does provide all the essential nutrients which can help the body against genetic complexities, early signs of aging, gut ailments, mental illness, skin problems, and other diseases. A group of experts from America studied this diet with respect to its fat content and its possible health impacts. And the results showed that the diet was effective in preventing cardiovascular diseases and increased the average age of life expectancies in the areas under study.

Since the major focus of the Mediterranean diet is on plant-based products, like grains, seeds, fruits, vegetables, oils, this is probably the true reason that it is full of important nutrients and devoid of fats or bad cholesterol and harmful toxins. From vegetables to grains and fruits, to dairy to meat and seafood, we can experience all using this diet, but it must be in a perfectly balanced proportion. Nutritionists and experts of the field from all around the globe have termed Mediterranean Food as 'The Best Ever," and we cannot deny the fact

Chapter 1. What is the Mediterranean diet?

The Mediterranean diet is primarily a heart-healthy eating plan based on the traditional food, drinks, meals, and recipes of the countries surrounding the Mediterranean Sea. To put it simply, the Mediterranean diet is adopting the Mediterranean cuisine and cooking style.

The Mediterranean diet is not a diet per se. You don't really go on a diet. Rather, the "diet" is a lifestyle that has been studied and noted to be as one of the healthiest in the world.

This eating plan or model does not only include healthy food, but it also involves physical activities, eating meals with family and friends, and drinking wine in moderation.

The Mediterranean diet is often also referred to as an eating strategy where you must "eat like a Greek", which stresses the following:

- Eating mainly vegetables, fruits, whole grains, nuts, legumes, and other plant-based foods
- Using olive oil, canola, and other healthy fats instead of butter
- Replacing salt with herbs and spices to flavor foods
- Eating red meat to not more than 1-2 times a month
- Eating fish and poultry at least 2 times a week
- Drinking red wine in moderation, and
- Getting plenty of physical activities or exercise

Just as important as eating healthy, the Mediterranean diet also emphasizes on preparing flavorful, delicious meals and dishes. If you are a novice in "eating like a Greek" you may think that the Mediterranean cooking-style is complex, but the beautiful fact about this diet, once you have begun the lifestyle changes and adopted the Mediterranean eating habit, this eating plan is very simple and fun. You'll find that there are tons of easy Mediterranean dishes and it does not take much to improve your health and lose weight and enjoy fabulous dishes right in the comforts of your own home.

Chapter 2. History and benefits of Mediterranean food

Origins of the Mediterranean Diet

The true origins of the Mediterranean diet stretch back into ancient times and are lost to history. However, we know that by the time of the ancient civilizations of Greece, Egypt, and Rome, what we now call the Mediterranean diet was commonly consumed in the Mediterranean basin. This diet was like what we have described so far, a diet based on bread, oil, and fish. The ancient Romans were known to consume large amounts of seafood, but as you might expect, the ability to consume fish and meat was closely tied to your social and economic status. As such, rural peoples consumed more bread, fruits, and vegetables relative to the elites who tended to eat large amounts of seafood. The consumption of red wine goes back to these ancient cultures as well. Wine was also consumed regularly by people of lesser means, and so bread, olive oil, and red wine was common in the diets of most people in the ancient Mediterranean world. Vegetables have always played a central role in the diets of Mediterranean people as well.

With the rise of Christianity, bread, olive oil, and red wine became identified with the Christian church, and this helped to preserve the traditional dietary patterns as the world rapidly changed after the fall of the western Roman empire.

Health Benefits of the Mediterranean Diet

We've already touched on the issue of health, noting that in the Mediterranean region, especially when adherence to traditional diets was widespread, the levels of chronic diseases were lower than in the United States, and still are in many cases. This is especially true when it comes to heart disease.

We've already mentioned that a scientist named Alan Keys discovered there was a relationship between cholesterol and heart disease. While we now know that the relationship is far more complicated than he imagined, and unfortunately his research results were misinterpreted to mean that people should follow strict low-fat diets, he also made some important observations about the Mediterranean countries.

Immediately after the war, Keys noted that people in war-torn areas of Italy and Greece seemed to be in pretty good health. At first, it was thought that obesity was rare because of the war, but it turns out that obesity was rare because of the diet most people were consuming at the time. There was also a documented lack of chronic "western" diseases.

Studies have compared following a high-fat Mediterranean diet with a traditional low-fat diet which is advised after a heart attack. It's been found that the Mediterranean diet rich in fat from seafood, olive oil, and nuts reduces the risk of a second heart attack by up to 70% as compared to following the traditional low-fat diet advocated by the American Heart Association and others.

Studies have also shown that following a Mediterranean diet reduces the risk of a heart attack at a level like that obtained by taking statin drugs.

Regular intake of Fruits and Vegetables lessens the chances of many diseases

We've already noted that fruits and vegetables help supply the body with vital potassium and magnesium. And in the case of leafy green vegetables, they also help supply calcium as well. These minerals help balance out the electrolytes in your body that control blood pressure and heart rhythm and keep muscles healthy as well. If you're getting a lot of muscle cramps, you might want to look at your consumption of potassium and magnesium.

However, the benefits of fruits and vegetables don't end there. Certainly, you're aware that they are packed with vitamins that are needed for health including vitamins A & C that will help you maintain a healthy immune system. Leafy green vegetables also contain vitamin K. There are two types of vitamin K, K1 is important for proper blood clotting, and K2 helps keep your arteries clear of calcium, which promotes heart health.

Fruits and vegetables also provide a large amount of dietary fiber. This helps keep your digestion regular, and it also helps you to maintain a healthy gut biome. Remember that your stomach and intestines are home for a large population of bacteria, most of which exist with us in a symbiotic relationship.

These bacteria help digest your food and keeping the right balance of healthy bacteria is important for health. One way to do this is by supplying them with the fibers contained in many fruits and vegetables.

Fruits and vegetables also provide many micronutrients, antioxidants, and so-called phytonutrients which can help reduce the risk of cancer.

In fact, leafy green vegetables play an important role in digestive health related to colon cancer. It's been found that in the absence of leafy greens, consuming red meat can cause the formation of certain cancer-causing substances in the colon that promote colon cancer. Eating red meat alone might cause benign polyps to turn into aggressive cancers. These substances are created by the bacteria in your intestines.

However, if you eat your steak with leafy green vegetables, this doesn't happen. The digestion of the leafy greens keeps the bacteria occupied and the cancer-causing substances they make during the digestion of red meat by itself aren't made. You might file this away in the steak + baked potato = bad and the steak + leafy greens = good files.

Health benefits of whole grains

Whole grain foods, as we've discussed before, provide a healthier way to get carbohydrates. Let's review – if you eat a meal high in sugar, you're going to get a blood sugar spike. The simpler the carbohydrates in the food you eat, the faster they're digested and the higher your blood sugar spike. Blood sugar spikes are bad for you, causing damage throughout the circulatory system.

Whole grain foods minimize blood sugar spikes. Simply put, it takes longer to break them down, so you'll get a longer but shallower and smoother rise in blood sugar. That's far healthier. Whole grain foods also provide a lot of important vitamins and minerals. For example, whole grain pasta made from wheat berries contains thiamin, niacin, riboflavin, B6, phosphorous, zinc, iron, magnesium, manganese, and potassium. In addition, it's rich in fiber.

To summarize, whole grain bread, pasta, and grains provide vitamins, minerals, and fiber, while providing liberal energy from carbohydrates without the blood sugar spikes.

Health benefits of olive oil

Olive oil might be considered as a magical elixir. While you probably don't want to drink a large glass and should watch it since fat packs a lot of calories in smaller amounts, it's something you want to use liberally, rather than sparingly in your diet.

In fact, olive oil-based diets have been directly compared to low-fat diets in large scientific studies. It's been demonstrated that following a Mediterranean style diet that uses liberal amounts of olive oil reduces the risk of contracting diabetes by 50% when compared to following a low-fat diet. The documented decrease in the risk of diabetes was found to be independent of other factors, like obesity at the beginning of the study or level of physical activity.

Studies in Europe have shown that liberal olive oil use helps reduce the incidence of stroke. In fact, one study found that people who used olive oil daily and in large amounts reduced their risk of stroke by up to 40%. Again, this reduction in risk was found to be independent of other factors like exercise level or body weight.

Olive oil also improves arterial function. As we age, our arteries don't work as well as they used to. They become stiffened and less responsive. Think of them as old pipes. However, researchers have found that olive oil helps to keep the arteries young and supple.

Another area where liberal use of olive oil helps is in preventing or even reversing metabolic syndrome. Remember that metabolic syndrome is characterized by a fat stomach or midsection. Inside the body, people with metabolic syndrome have high blood sugar, high blood sugar, high levels of bad LDL cholesterol and triglycerides, and low HDL or "good" cholesterol. Large studies have shown that liberal use of olive oil can have moderate but significant effects in reducing the waistline, reducing blood pressure, reducing bad cholesterol, reducing triglycerides, and raising HDL cholesterol. This was modest but independent of other factors. In other words, it can help a person overcome metabolic syndrome when used in combination with other lifestyle changes.

Olive oil may also protect against the development of certain cancers. Olive oil contains a phytonutrient named oleocanthal, which helps reduce inflammation in the body. By reducing inflammation, it helps reduce the risk of certain cancers including breast and prostate cancer. Inflammation is also an important factor in the development of heart disease, by the way.

Olive oil contains many fats of all types, including 11% saturated fat. However, it's about 73% monounsaturated fat. Monounsaturated fats have been shown to reduce bad cholesterol and lower triglycerides – so consumption of liberal amounts of monounsaturated fats like olive oil can significantly lessen the chances of a heart attack and stroke.

Monounsaturated fats also slightly lower blood pressure. They also reduce insulin resistance (see chapter one), making the body more efficient when processing carbohydrates.

An inflammatory marker that is associated with elevated heart attack risk is called C - reactive protein, or CRP. It's been found that consuming a diet rich in olive oil can reduce CRP by about 15%. And the more olive oil consumed, the more CRP was reduced. By reducing the level of inflammation, you can reduce the chance of sudden death from a heart attack.

Olive oil also supplies some important vitamins – specifically it provides vitamins E and K.

Experts recommend consuming extra-virgin olive oil.

Health benefits of consuming fatty fish

Fatty fish is consumed throughout the Mediterranean region, where sardines, mackerel, tuna, and swordfish are very popular. In addition, people consume many anchovies.

Eating fatty fish has two benefits. By eating fish rather than beef, you're reducing the amount of saturated fat in the diet. Remember that while saturated fat in and of itself isn't necessarily bad, it does raise LDL cholesterol. All else being equal, that's not a good thing (although as we discussed in the previous chapter, the effect can be harmless if your triglycerides are low).

The main benefit of eating fish, however, is that fatty fish supplies the body with omega-3 fatty acids. Omega-3 fats have several health benefits.

- They stabilize heart rhythms.
- They reduce inflammation.
- They reduce triglycerides.

Overall, people who consume fatty fish in large amounts have far lower rates of heart disease and stroke. People expected that they could duplicate these benefits by packaging fish oil in pills, but that hasn't turned out to be as fruitful as people originally hoped. It may be that the omega-3 oils in pills could reduce the incidence of heart disease, but they aren't given in the correct dosages. In any case, eating fatty fish at least twice a week has been conclusively demonstrated significantly lessen the risk of heart disease.

One area where there has been a success with omega-3s in pill form is using prescription strength fish oils to lower triglycerides. It may be that only people with high triglycerides benefit from fish oil and they need it in high doses, but that isn't clear at this time. However, a can of sardines, mackerel or anchovies every day can lower triglycerides just as well as the prescription fish oil capsules.

By lowering inflammation, fish oils may also reduce the risk of certain cancers as well.

It may be the case that consuming the whole fish is important for fighting heart disease, rather than getting the oils in pill form that may leave out some unknown cofactors.

Fatty fish also helps you feel satiated with a meal in a way that lean meat will not. Although salmon isn't consumed in the Mediterranean region, it's a perfectly acceptable food since it has a similar nutrition profile to sardines and mackerel.

Health benefits of nuts and seeds

Nuts and seeds provide many key nutrients that lead to a healthier lifestyle. At the forefront of the many health benefits of nuts is that like olive oil, nuts contain a large amount of healthy monounsaturated fat. Nuts also contain many minerals they are good sources of potassium, magnesium, and some nuts have significant amounts of calcium. If you recall from our discussion about the origins of the DASH diet, nuts have many of the nutrients that are necessary to bring your electrolytes into balance and achieve or maintain healthy blood pressure levels. The monounsaturated fats found in nuts also contribute to health in the same way that olive oil contributes to health, helping to reduce inflammation and improve blood lipids. Nuts are high in calories, so they are restricted on the DASH diet. On a Mediterranean diet, you can eat as many as you like if you don't overeat (eat past being full).

The monounsaturated fats found in nuts help lower bad cholesterol. In fact, studies that examine the consumption of nuts have found that people who eat nuts daily have lower rates of heart attacks and stroke.

In addition to supplying vital minerals, nuts are also a good source of vitamin E. Nuts also supply large amounts of dietary fiber, making them a good adjunct to whole grains, fruits, and vegetables.

Some nuts, walnuts, contain a form of omega-3 fatty acids. However, the form of omega-3 fats in plants and in this case, nuts is not utilized by the body as easily as the omega-3 fats found in fish. When seeking omega-3 fat you should eat fish, but nuts should be consumed for other reasons.

Nuts also contain an amino acid called L-arginine. This little fellow plays an important role in keeping your arteries healthy, they make your arteries more flexible and lowers the occurrence of blot clots.

Good nuts to consume include almonds, walnuts, pecans, pistachios, and macadamia nuts. Cashews have a bit of carb in them so you might try consuming them less often. In any case, snacking on a handful of nuts daily is a good way to maintain lifelong health. You can also use nuts and seeds to top off salads or include them with stir-fried vegetables.

The Benefits of Red Wine

Red wine contains many phytonutrients and antioxidants. Moderate consumption of alcohol has even been shown to increase lifespan. This effect is known as the "J-curve". That is, relative to moderate drinkers, those who don't drink at all have an elevated risk of death. While there is a sweet spot where you get the most benefit, drinking beyond that begins to increase your risk of cancer and other problems like stomach ulcers and bleeding. So, people who are heavy drinkers tend to have a much higher risk of death as compared to people who don't drink or moderate drinkers.

Red wine has also been associated with a lower risk of a heart attack. France has low death rates due to heart disease despite high fat consumption, and red wine is believed to be the reason. It might raise HDL cholesterol slightly and thins the blood a little bit that may reduce the risk of blood clots that could cause heart attacks and strokes (when consumed in moderation, heavy drinkers have a serious risk of internal bleeding).

Some studies have also shown that red wine can help maintain healthy levels of omega-3 fats and maintain healthy blood sugar levels.

However, it almost goes without saying – if you don't drink now, you're probably better off not starting. And if you have problems controlling alcohol consumption you probably shouldn't drink at all.

After the fall of the western Roman empire, the Arab world had a large influence on dietary patterns in the Mediterranean basin. This led to an increased prominence of fruit consumption, including citrus fruits like oranges and lemons. Islamic culture also had an influence on the use of spices by Europeans, but this was generally confined to the upper socioeconomic classes. However, it set the stage for the development of the modern idea of the Mediterranean diet.

The voyages of Europeans to the Americas also had a large influence on the development of the Mediterranean diet. This led to the introduction of the tomato and potato into European diets, and the tomato has become associated with the Mediterranean style of eating. This red fruit is extremely nutritious and livens up salads in addition to being well-suited for making tasty sauces.

Chapter 3. How to Lose Weight by Eating Healthy?

If you are wondering about the initial loss of weight on any diet, it really did happen. It wasn't the handiwork of a faulty scale. There is no doubt that you can lose some weight when you get on any kind of diet. Weight loss is real.

However, it is unsustainable.

When you get on a diet, you restrict the number of calories ingested by you.

This means that if you are running your body on 2500 calories per day, your body gets used to the process. It runs all the metabolic functions in a manner that they consume around these many calories. Yes, it is a fact that your body burns several calories every day.

The number of calories burned by your body in the idle state when you do nothing for a day besides respirating is called the Basal Metabolic Rate or BMR. It is the minimum calorie requirement of your body to carry out all the metabolic functions smoothly. Diets try to bring down the number of calories below the BMR.

For instance, if your BMR is 2000, diets would try to lower your intake to 1500 calories. This means that you would be supplying 500 calories lesser than required and expect your body to fulfill the shortage from the stored fat. This is the next logical thing to do. However, it is not a prudent thing to do, and your body knows better.

It has passed through millions of years of the evolutionary process, and it knows that if it starts depleting its stores at the drop of a hat, there might not be any reserve needed when you really need it to survive in the actual famine periods. Hence, it doesn't start burning the fat deposits. It reduces its energy expenditure and starts rationing in the lavish use of energy.

A large part of the energy in our body is spent on making it feel comfortable. Fending off heat and cold is a part of the process. To provide insulation against heat and cold, our body stores water in large quantity. Keeping that water hot or cold as per the season's requirement consumes a lot of energy. As soon as your body senses substantial energy shortage, it starts dumping excess water. It knows that staying alive for longer is more important than feeling warm or comfy.

This is a reason, people who begin dieting become very sensitive to hot and cold. They have a shaky feeling within. They also become very cranky as the body is actively looking for the restoration of the usual supply of energy.

So, if you initially lost weight on a diet, there is no surprise. However, such people would observe that although their weight goes down on the scale, the size of their waist, hips, thighs, and belly remains the same. The reason is simple; fat loss is not taking place at all.

Such loss of weight is never permanent. Our body is very clever. It soon devises ways to compensate for the energy shortage and makes itself more efficient. It would then again regain the lost weight.

Not only this, but the weight relapse would also happen when a person gets off a diet. Then, the body again starts getting the usual energy intake, and hence, there is no need to do rationing. People also indulge in binge eating after diets as they had been starving mentally and emotionally. This leads to binge eating and causes weight gain.

This kind of weight loss is not sustainable because the real fat burning process never started at all.

Chapter 4. First two days of detoxification from junk food

Detoxification has gained a nasty reputation these past years. While there are countless detox products that claim to be "the only weight loss solution", we all know by now that liquid diets simply don't work on their own. You may experience a sudden drop in your weight in the beginning, but usually that also means a drop-in nutrients and energy. It may help you get a jumpstart in your weight loss journey, but it also gets difficult to sustain along the way. A tea detox, or teat ox as most celebrities like to call it, is a much healthier approach to detoxifying your body. Instead of replacing full meals with a liquid drink, you only need to add a few cups of herbal tea to your already existing, nourishing diet. This means that you can still have all the fruits and vegetables you want even while you're trying to cleanse your body of all the harmful toxins that are trapped in your bloodstream.

Because detox tea is so easy to incorporate into anyone's lifestyle, it's no wonder that countless celebrities now swear by its amazing effects. But what is it about tea that makes it the best weight loss solution on the market today? Here's how tea can help you get started on a healthier and happier way of life.

According to a 2013 study conducted by American researches, going a tea-drinking binge has a wide array of benefits that covers almost every area of the human body. From lowering your risk of stroke, to increasing mental performance, tea is packed with catechins that can help you keep your energy level up even with less calorie consumption. This is probably the main reason why tea drinkers cope better both physically and emotionally when they make changes to their lifestyle.

High quality teas, both green and black, are rich in antioxidants that can help boost the body's natural cleansing ability. Antioxidants play a crucial role in the detoxification process because it reduces oxidative stress levels significantly and gets rid of free radicals from the body. While drinking tea alone isn't enough to get the job done, it can make the detoxifying process much easier for the body. It's considered to be harmless compared to many detox products that are designed to just mess up the body's natural cycle.

And because there are teas specifically blended with additional ingredients like lemongrass, dandelion, and even milk thistle, you're sure to get more benefits from doing a teat ox than a traditional detox. You can choose the perfect tea blend that will help you meet your specific health and fitness goals. If you're looking for a detoxifying drink that will alleviate stress on the liver, an herbal infusion with ginger for example can clean your bloodstream more efficiently. It's just a matter of finding the right tea blend that will suit not just your mood or taste, but also complement your body system.

But keep in mind that not teat ox teas are created equally. There are some that contain a very powerful detoxifying, but dangerous ingredient called Senna. Senna is an herbal laxative that stimulates the intestines to purge its contents. While this ingredient can be helpful on the body for a short amount of time, taking too much Senna for too long can have devastating effects on the digestive system. It can cause electrolyte balance which if you're not careful, can lead to dehydration. If you feel constipated, taking Senna tea for a few nights can be helpful but don't let it become your everyday cup.

So, when's the best time to take your cup of tea? Health experts believe that you can take it whenever you feel like it, if you make the effort to drink more water throughout the day since most tea blends contain caffeine. But for tea drinkers who can only stomach 1-2 cups of tea per day, it's best to take your first cup once you wake up, and your second one before you prep for bed. This way, you'll get your dose of antioxidants without having to make any major changes in your daily routine.

Whatever teat ox blend you choose, make sure that you eat a healthy diet with it. Going on a tea detox can only do so much without the help of a proper diet plan. If you want to detoxify your body, you need to make that life changing decision to cut out processed foods from your food plan. You need to feed your body with fruits, vegetables, and whole grains in order to enhance your digestive system's natural cycle. Once you start getting the hang of eating clean, detoxifying your system is going to be a breeze.

Chapter 5. 4-week meal plan

DAY	BREAKFAST	MAINS	DESSERTS
1.	Cranberry Granola Bars	Garlicky Roasted Chicken	Greek Yogurt Frosted Zucchini Cupcakes with
2.	Breakfast Kale Frittata	Tomato Roasted Feta	Apricots and Mascarpone Cream
3.	Homemade Granola Bowl	Feta Stuffed Pork Chops	Minty Orange Greek Yogurt
4.	Mushroom Frittata	Creamy Smoked Salmon Pasta	Apricot Almond Dips
5.	Apple muffins	Roasted Chicken in Salt Crust	Rustic Raspberry and Fig Mini Crostatas
6.	Creamy Millet	Roasted Red Bell Pepper Chicken Stew	Pasta Flora or Greek Tart with Apricot Jam

7.	Rice Pudding	Potato Salmon Casserole	Frozen Strawberry Greek Yogurt
8.	Spiced Morning Omelet	Mediterranean Roasted Lamb and Sweet Potatoes	Orange-Sesame Almond Toiles
9.	Green Beans and Eggs	Quick Zucchini Stew	Kataifi
10.	Sweet Oatmeal	Chicken Meatballs in Herbed Tomato Sauce	Hazelnut-Orange Olive Oil Cookies
11.	Quinoa Bowl	Chicken Zucchini Ragout	Greek Cheesecake
12.	Morning Egg Sandwiches	Grilled Pesto Salmon	Phyllo Cups
13.	Watermelon Salad	Barley with Perfect Roasted Vegetables	Poached Cherries

14.	Cucumber and Avocado Salad	Couscous Casserole with Peppers and Goat Cheese	Mediterranean Bread Pudding
15.	Spinach and Berry Smoothie	Savory Oatmeal with Mozzarella Cheese	Mediterranean Cheesecake
16.	Zucchini Breakfast Salad	Rosemary Roasted New Potatoes	Kale Wraps with Apple and Chicken
17.	Simple Basil Tomato Mix	Slow Cooked Cod Stew	Greek Style Nachos
18.	Quinoa and Spinach Breakfast Salad	Mediterranean Tuna Steaks	Lemon Cauliflower Florets
19.	Carrots Breakfast Mix	Herbed Chicken Stew	Italian Style Potato Fries
20.	Zucchini and Sprout Breakfast Mix	Pesto Braised Roasted Leg	Sweet Potato Fries
21.	Breakfast Corn Salad	Pear Braised Pork	Grilled Tempeh Sticks

22.	Italian Breakfast Salad	Saucy Kidney Beans with Kale	Glazed Mediterranean Puffy Fig
23.	Cranberry Granola Bars	Black-Eyed Pea Bowl with Scallions	Mediterranean Stuffed Custard Pancakes
24.	Breakfast Kale Frittata	Vermicelli with Beans and Lemon Crema Fresca	Mascarpone and Ricotta Stuffed Dates
25.	Homemade Granola Bowl	Orzo Pilaf with Herbs	Mediterranean Stuffed Dates
26.	Mushroom Frittata	Turkish Pilaf with Roasted Chickpeas	Watermelon-Strawberry Rosewater Yogurt Panna Cotta
27.	Apple muffins	Cornbread Squares with Vegetables	Sushi Appetizer
28.	Creamy Millet		Tuna Salad in Lettuce Cups
29.	Rice Pudding		Rice Burgers

30.	Spiced Morning Omelet		Tzatziki

Chapter 6. Breakfast

Morning Egg Sandwiches

Servings: 4
Preparation Time: 10 minutes
Cooking Time: 10 minutes
Ingredients
5 oz whole grain bread
1 tablespoon sunflower seeds butter
¼ teaspoon salt
1 avocado, pitted
4 eggs
Directions:
Slice the bread into 8 slices.
Preheat a skillet and add the sunflower seeds butter and melt it well.
Beat the eggs in the skillet and sprinkle them with the salt.
Chop the avocado into the medium cubes and mash it well.
Spread the bread slices with the avocado mash.
When the eggs are cooked, cool them a little and place on top of the bread slices to make the sandwiches.
Serve the dish immediately.
Nutrition: calories: 275, fat: 17.7g, total carbs: 21.7g, sugars: 3.2g, protein: 11.1g

Quinoa Bowl

Servings: 6
Preparation Time: 10 minutes
Cooking Time: 15 minutes
Ingredients
2 cups quinoa
1 cup blueberries
1 cup coconut milk, unsweetened
2 cups water

2 tablespoons almonds
1 teaspoon pistachio
2 tablespoons honey
Directions:
Combine the coconut milk and water together in the saucepan and stir the liquid well.
Add the quinoa and close the lid.
Cook the mixture on medium heat for 5 minutes.
Wash the blueberries carefully and add them to the quinoa mixture.
Stir it carefully and continue to cook.
Combine the pistachio and almonds together and crush the nuts.
Sprinkle the quinoa with the crushed nuts and cook the mixture for 3 minutes more.
Add honey and stir the mixture carefully until honey has dissolved.
Transfer to serving bowls and enjoy.
 Enjoy!
Nutrition: calories: 348, fat: 14.1g, total carbs: 48.3g, sugars: 9.6g, protein: 9.6g

Sweet Oatmeal

Servings: 3
Preparation Time: 5 minutes
Cooking Time: 10 minutes
Ingredients
1 cup oatmeal
5 apricots
1 tablespoon honey
1 cup coconut milk, unsweetened
1 teaspoon cashew butter
¼ teaspoon salt
½ cup water
Directions:

Combine the coconut milk and oatmeal together in the saucepan and stir the mixture.
Add the water and stir it again. Sprinkle the mixture with the salt and close the lid.
Cook the oatmeal on medium heat for 10 minutes.
Meanwhile, chop the apricots into tiny pieces and combine the chopped fruit with the honey.
When the oatmeal is cooked, add cashew butter and fruit mixture.
Stir carefully and transfer to serving bowls.
Serve immediately.
Nutrition: calories: 336, fat: 21.2g, total carbs: 35.1g, sugars: 14.0g, protein: 6.2g

Green Beans and Eggs

Servings: 2
Preparation Time: 10 minutes
Cooking Time: 15 minutes
Ingredients
½ cup green beans
¼ teaspoon salt
5 eggs
1/3 cup skim milk
1 bell pepper, seeds removed
1 teaspoon olive oil
Directions:
Slice the bell pepper and combine it with the green beans.
Pour the olive oil in a skillet and transfer the vegetable mixture to the skillet.
Cook on medium heat for 3 minutes, stirring frequently.
Meanwhile, beat the eggs in a mixing bowl.
Sprinkle the egg mixture with the salt and add skim milk. Whisk well.
Pour the egg mixture over the vegetable mixture and cook for 3 minutes on medium heat.

Stir the mixture carefully so that the eggs and vegetables are well combined.
Cook for 4 minutes more.
Stir again and close the lid.
Cook the scrambled eggs for 5 minutes more.
Stir the mixture again.
Serve it.
Nutrition: calories: 231, fat: 13.4g, total carbs: 9.3g, sugars: 6.2g, protein: 16.3g

Spiced Morning Omelet

Servings: 3
Preparation Time: 10 minutes
Cooking Time: 15 minutes
Ingredients
7 eggs
1/3 cup skim milk
3 garlic cloves
¼ teaspoon nutmeg
¼ teaspoon ground ginger
1 teaspoon cilantro
1 teaspoon olive oil
1 tablespoon chives
1 teaspoon turmeric
Directions:
Beat the eggs in a mixing bowl.
Add the skim milk and whisk again.
Sprinkle the egg mixture with the nutmeg, ground ginger, cilantro, and turmeric.
Peel the garlic cloves and mince them.
Chop the chives and combine with the minced garlic.
Add the herb mixture to the eggs and stir it again.
Preheat a skillet well and pour in the olive oil.
Preheat the olive oil over medium heat and then pour the egg mixture into the pan.

Close the lid and cook the omelet for 15 minutes.
When the dish is cooked, cool slightly and cut into the serving portions.
Serve it.
Nutrition: calories: 179, fat: 12.0g, total carbs: 3.8g, sugars: 2.2g, protein: 14.1g

Rice Pudding

Servings: 5
Preparation Time: 5 minutes
Cooking Time: 15 minutes
Ingredients
1 cup brown rice
2 cups coconut milk, unsweetened
1 teaspoon cinnamon
1 teaspoon ginger
1/3 teaspoon thyme
1/3 cup almonds
2 tablespoon honey
1 teaspoon lemon zest
Directions:
Pour the coconut milk into a saucepan and heat over medium.
Add the brown rice and stir the mixture carefully.
Close the lid and cook the brown rice over medium heat for 10 minutes.
Meanwhile, crush the almonds and combine them with the lemon zest, thyme, ginger, and cinnamon.
Sprinkle the brown rice with the almond mixture and stir it carefully.
Close the lid and cook the dish for 5 minutes.
When the pudding is cooked, remove it from the saucepan and transfer to a big bowl.
Add the honey and stir the pudding.
Serve it immediately.

Nutrition: calories: 423, fat: 27.1g, total carbs: 43.3g, sugars: 10.4g, protein: 6.5g

Creamy Millet

Servings: 8
Preparation Time: 10 minutes
Cooking Time: 15 minutes
Ingredients
2 cups millet
1 cup almond milk, unsweetened
1 cup water
1 cup coconut milk, unsweetened
1 teaspoon cinnamon
½ teaspoon ground ginger
¼ teaspoon salt
1 tablespoon chia seeds
1 tablespoon cashew butter
4 oz Parmesan cheese, grated
Directions:
Combine the coconut milk, almond milk, and water together in the saucepan.
Stir the liquid gently and add millet.
Mix carefully and close the lid.
Cook the millet on the medium heat for 5 minutes.
Sprinkle the porridge with the cinnamon, ground ginger, salt, and chia seeds.
Stir the mixture carefully with a spoon and continue to cook on medium heat for 5 minutes more.
Add the cashew butter and cook the millet for 5 minutes.
Remove the millet from the heat and transfer it to serving bowls.
Sprinkle the dish with the grated cheese.
 Serve it.
Nutrition: calories: 384, fat: 19.8g, total carbs: 42.9g, sugars: 3.6g, protein: 11.7g

Apple muffins

Servings: 5
Preparation Time: 10 minutes
Cooking Time: 15 minutes
Ingredients
2 eggs
1 cup oat flour
½ teaspoon salt
2 tablespoon stevia
3 apples, washed and peeled
½ cup skim milk
1 tablespoon olive oil
½ teaspoon baking soda
1 teaspoon apple cider vinegar
Directions:
Beat the eggs in the mixing bowl and whisk them well.
Add the skim milk, salt, baking soda, stevia, and apple cider vinegar.
Stir the mixture carefully.
Grate the apples and add the grated mixture in the egg mixture.
Stir it carefully and add the oat flour.
Add the olive oil and blend into a smooth batter
Preheat the oven to 350 F.
Fill each muffin form halfway with the batter and place the muffins in the oven.
Cook the dish for 15 minutes.
 Remove the cooked muffins from the oven.
 Cool the cooked muffins well and serve them.
Nutrition: calories: 200, fat: 6.0g, total carbs: 32.4g, sugars: 15.3g, protein: 11.7g

Mushroom Frittata

Servings: 5
Preparation Time: 10 minutes
Cooking Time: 20 minutes
Ingredients
8 oz shiitake mushrooms
1 teaspoon salt
1 cup broccoli
7 eggs
5 oz Parmesan cheese
1 tablespoon olive oil
½ teaspoon ground ginger
5 garlic cloves
1 teaspoon oregano
1 teaspoon basil
1 teaspoon cilantro
½ cup low-fat milk
Directions:
Wash the shiitake mushrooms well and chop them.
Chop the broccoli and combine it with the mushrooms in a mixing bowl.
In a separate bowl, beat the eggs.
Sprinkle the egg mixture with the cilantro, basil, oregano, and ground ginger. Stir it well.
Add the low-fat milk and broccoli. Stir the egg mixture well.
Peel the garlic cloves and mince them.
Add minced garlic in the egg mixture and stir it gently.
Preheat the oven to 350 F.
Spray a deep pan with olive oil
Pour the egg mixture into the pan and place it in the preheated oven.
Cook the frittata for 20 minutes.
When the dish is cooked, remove it from the oven and cool slightly.
Serve the frittata immediately.
Nutrition: calories: 250, fat: 15.5g, total carbs: 11.5g, sugars: 3.7g, protein: 19.2g

Homemade Granola Bowl

.
Servings: 6
Preparation Time: 10 minutes
Cooking Time: 20 minutes
Ingredients
3 tablespoons pumpkin seeds
1 tablespoon coconut oil
1 teaspoon sunflower seeds
¼ cup almonds
1 cup raw oats
3 tablespoons sesame seeds
5 tablespoons honey
2 cups almond milk, unsweetened
Directions:
Combine the pumpkin seeds, sunflower seeds, almonds, and sesame seeds together.
Crush the mixture well and add raw oats.
Add the honey and coconut oil.
Stir the mixture carefully until you get a smooth mix.
Preheat the oven to 350 F.
Cover the tray with parchment and transfer the seed mixture onto the tray. Flatten it well.
Put the tray in the preheated oven and cook it for 20 minutes.
When the mixture is cooked, remove it from the oven and chill well.
Separate the mixture into small pieces and put in serving bowls.
Add the almond milk and mix up the dish.
Serve it.
Nutrition: calories: 381, fat: 28.5g, total carbs: 30.8g, sugars: 17.4g, protein: 6.4g

Breakfast Kale Frittata

Preparation time: 10 minutes
Cooking time: 30 minutes
Servings: 4
Ingredients:

6 kale stalks, chopped
1 small sweet onion, chopped
1 small broccoli head, florets separated
2 garlic cloves, minced
Salt and black pepper to the taste
4 eggs
1 tablespoon olive oil

Directions:
Heat up a pan with the oil over medium-high heat, add the onion, stir and cook for 4-5 minutes. Add the garlic, broccoli and kale, toss and cook for 5 minutes more. Add the eggs, salt and pepper and mix. Place in the oven and bake at 380 degrees F for 20 minutes. Slice and serve for breakfast.
Enjoy!
Nutrition: calories 214, fat 7, fiber 2, carbs 12, protein 8

Cranberry Granola Bars

Preparation time: 2 hours
Cooking time: 0 minutes
Servings: 4
Ingredients:

2 cups walnuts, toasted
1 cup dates, pitted
3 tablespoons water
¾ cup cranberries, dried, no added sugar

2 cups desiccated coconut, unsweetened

Directions:
In your food processor, mix dates with coconut, cranberries, water and walnuts. Pulse well then spread the mix into a lined baking dish. Press well into the dish and keep in the fridge for 2 hours then cut into bars and serve.
Enjoy!
Nutrition: calories 476, fat 40, fiber 9, carbs 33, protein 6

Spinach and Berry Smoothie

Preparation time: 10 minutes
Cooking time: 0 minutes
Servings: 2
Ingredients:

1 cup blackberries
1 avocado, pitted, peeled and chopped
1 banana, peeled and roughly chopped
1 cup baby spinach
1 tablespoon hemp seeds
1 cup water
½ cup almond milk, unsweetened

Directions:
In your blender, mix the berries with the avocado, banana, spinach, hemp seeds, water and almond milk. Pulse well, divide into 2 glasses and serve for breakfast.
Enjoy!
Nutrition: calories 160, fat 3, fiber 4, carbs 6, protein 3

Zucchini Breakfast Salad

Preparation time: 10 minutes
Cooking time: 0 minutes
Servings: 4
Ingredients:

2 zucchinis, spiralized
1 cup beets, baked, peeled and grated
½ bunch kale, chopped
2 tablespoons olive oil
For the tahini sauce:
1 tablespoon maple syrup

Juice of 1 lime
¼ inch fresh ginger, grated
1/3 cup sesame seed paste

Directions:
In a salad bowl, mix the zucchinis with the beets, kale and oil. In another small bowl, whisk the maple syrup with lime juice, ginger and sesame paste. Pour the dressing over the salad, toss and serve it for breakfast.
Enjoy!
Nutrition: calories 183, fat 3, fiber 2, carbs 7, protein 9

Quinoa and Spinach Breakfast Salad

Preparation time: 10 minutes
Cooking time: 0 minutes
Servings: 2
Ingredients:

16 ounces quinoa, cooked
1 handful raisins
1 handful baby spinach leaves
1 tablespoon maple syrup
½ tablespoon lemon juice
4 tablespoons olive oil
1 teaspoon ground cumin
A pinch of sea salt and black pepper
½ teaspoon chili flakes

Directions:
In a bowl, mix the quinoa with the spinach, raisins, cumin, salt and pepper and toss. Add the maple syrup, lemon juice, oil and chili flakes and toss then serve for breakfast.

Enjoy!
Nutrition: calories 170, fat 3, fiber 6, carbs 8, protein 5

Carrots Breakfast Mix

Preparation time: 10 minutes
Cooking time: 0 minutes
Servings: 4
Ingredients:

1½ tablespoon maple syrup
1 teaspoon olive oil
1 tablespoon chopped walnuts
1 onion, chopped
4 cups shredded carrots
1 tablespoon curry powder
¼ teaspoon ground turmeric
Black pepper to the taste
2 tablespoons sesame seed paste
¼ cup lemon juice
½ cup chopped parsley

Directions:
In a salad bowl, mix the onion with the carrots, turmeric, curry powder, black pepper, lemon juice and parsley. Add the maple syrup, oil, walnuts and sesame seed paste. toss well and serve for breakfast.
Enjoy!
Nutrition: calories 150, fat 3, fiber 2, carbs 6, protein 8

Italian Breakfast Salad

Preparation time: 10 minutes
Cooking time: 0 minutes
Servings: 4
Ingredients:

1 handful kalamata olives, pitted and sliced
1 cup cherry tomatoes, halved
1½ cucumbers, sliced
1 red onion, chopped
2 tablespoons chopped oregano
1 tablespoon chopped mint
For the salad dressing:
2 tablespoons balsamic vinegar
¼ cup olive oil
1 garlic clove, minced
2 teaspoons dried Italian herbs
A pinch of salt and black pepper

Directions:
In a salad bowl, toss together the olives with the tomatoes, cucumbers, onion, mint and oregano. In a smaller bowl, whisk the vinegar with the oil, garlic, Italian herbs, salt and pepper. Pour the dressing over the salad, toss and serve for breakfast. Enjoy!
Nutrition: calories 191, fat 10, fiber 3, carbs 13, protein 1

Zucchini and Sprout Breakfast Mix

Preparation time: 10 minutes
Cooking time: 0 minutes
Servings: 4
Ingredients:

2 zucchinis, spiralized
2 cups bean sprouts
4 green onions, chopped
1 red bell pepper, chopped
Juice of 1 lime
1 tablespoon olive oil
½ cup chopped cilantro
¾ cup almonds chopped
A pinch of salt and black pepper

Directions:
In a salad bowl, toss together the zucchinis with the bean sprouts, green onions, bell pepper, cilantro, almonds, salt, pepper, lime juice and oil. Serve for breakfast.
Nutrition: calories 140, fat 4, fiber 2, carbs 7, protein 8

Breakfast Corn Salad

Preparation time: 10 minutes
Cooking time: 0 minutes
Servings: 4
Ingredients:

2 avocados, pitted, peeled and cubed
1-pint mixed cherry tomatoes, halved
2 cups fresh corn kernels
1 red onion, chopped
For the salad dressing:
2 tablespoons olive oil
1 tablespoon lime juice
½ teaspoon grated lime zest
A pinch of salt and black pepper
¼ cup chopped cilantro

Directions:
In a salad bowl, mix the avocados with the tomatoes, corn and onion. Add the oil, lime juice, lime zest, salt, pepper and the cilantro, toss and serve for breakfast.
Nutrition: calories 140, fat 3, fiber 2, carbs 6, protein 9

Simple Basil Tomato Mix

Preparation time: 10 minutes
Cooking time: 0 minutes
Servings: 6
Ingredients:

½ cup extra-virgin olive oil
1 cucumber, chopped
2 pints colored cherry tomatoes, halved
Salt and black pepper to the taste

1 red onion, chopped
3 tablespoons red vinegar
1 garlic clove, minced
1 bunch basil, roughly chopped

Directions:
In a salad bowl, toss together the cucumber with the tomatoes, onion, salt, pepper, oil, vinegar, basil and garlic. Serve for breakfast.
Enjoy!
Nutrition: calories 100, fat 1, fiber 2, carbs 2, protein 6

Cucumber and Avocado Salad

Preparation time: 10 minutes
Cooking time: 0 minutes
Servings: 4
Ingredients:

1-pound cucumbers, chopped
2 avocados, pitted and chopped
1 small red onion, thinly sliced
2 tablespoons olive oil
2 tablespoons lemon juice
¼ cup chopped parsley
A pinch of salt and black pepper

Directions:
In a salad bowl, mix the cucumbers with the avocados, onion, oil, lemon juice, parsley, salt and pepper. Serve for breakfast.
Enjoy!
Nutrition: calories 120, fat 2, fiber 2, carbs 3, protein 4

Watermelon Salad

Preparation time: 10 minutes
Cooking time: 0 minutes
Servings: 2
Ingredients:

½ teaspoon agave nectar
2 tablespoons lemon juice
1 tablespoon extra-virgin olive oil
1 jalapeno, seeded and chopped
12 ounces watermelon, chopped
1 red onion, thinly sliced
½ cup chopped basil leaves
2 cups baby arugula

Directions:
In a bowl, toss together the watermelon with the jalapeno, onion, basil, arugula, oil, agave nectar, lemon juice and oil. Serve for breakfast.
Nutrition: calories 128, fat 8, fiber 2, carbs 16, protein 2

Chapter 7. Lunch

Quick Zucchini Stew

Preparation time:5 minutes
Cooking time:40 minutes
Servings: 4
Ingredients:
2 tablespoons olive oil
1 shallot, chopped
4 garlic cloves, minced
1 red pepper, sliced
2 tomatoes, cubed
½ cup tomato juice
½ cup vegetable stock
4 zucchinis, cubed
1 tablespoon all-purpose flour
Salt and pepper to taste
2 tablespoons chopped dill
Directions:
Heat the oil in a skillet and stir in the shallot and garlic. Cook for 5 minutes then add the red pepper, tomatoes, juice, stock and zucchinis.
Season with salt and pepper and cook on low heat for 15 minutes.
Sprinkle in the flour and cook for another 5 minutes.
Add the dill and mix well then remove off heat.
Serve the stew warm and fresh.
Nutrition: Calories:135, Fat:7.7g, Protein:4.2g. Carbohydrates:15.9g

Chicken Zucchini Ragout

Preparation time:5 minutes
Cooking time:2 ½ minutes
Servings: 8
Ingredients:
3 tablespoons olive oil

2 pounds ground chicken
4 garlic cloves, minced
3 shallots, chopped
2 carrots, sliced
3 zucchinis, cubed
1 can diced tomatoes
1 cup chicken stock
2 tablespoons tomato paste
1 bay leaf
1 thyme sprig
Salt and pepper to taste
Directions:
Heat the oil in a heavy saucepan and stir in the ground chicken.
Cook for 5 minutes then add the garlic and shallots.
Cook for 5 minutes then stir in the rest of the ingredients.
Add enough salt and pepper then lower the heat and cook for 30 minutes.
Serve the ragout warm and fresh.
Nutrition: Calories:290, Fat:13.9g, Protein:34.4g, Carbohydrates:6.4g

Chicken Meatballs in Herbed Tomato Sauce

Preparation time:5 minutes
Cooking time:1 ½ hours
Servings: 8
Ingredients:
2 pounds ground chicken
4 garlic cloves, minced
1 shallot, chopped
2 tablespoons chopped parsley
1 teaspoon dried oregano
1 egg
Salt and pepper to taste
2 tablespoons olive oil
1 sweet onion, chopped
2 cups tomato sauce
1 rosemary sprig

1 thyme sprig
1 oregano sprig
½ cup dry red wine
1 teaspoon sherry vinegar
Directions:
Mix the ground chicken, garlic, shallot, parsley, oregano and egg in a bowl. Add salt and pepper and mix well. Form small meatballs and place them aside.
Heat the oil in a skillet and add the onion. Cook for 2 minutes then stir in the tomato sauce, rosemary, thyme, oregano, red wine and vinegar.
Bring to a boil then place the meatballs in the hot sauce. Cook on low heat for 20 minutes.
Serve the meatballs and the sauce warm and fresh.
Nutrition: Calories:290, Fat:12.6g, Protein:34.7g, Carbohydrates:5.9g

Garlicky Roasted Chicken

Preparation time:5 minutes
Cooking time:2 ½ hours
Servings: 8
Ingredients:
1 whole chicken
8 garlic cloves, minced
2 tablespoons chopped rosemary
1 teaspoon dried basil
¼ cup butter, softened
Salt and pepper to taste
Directions:
Mix the garlic, rosemary, basil, butter, salt and pepper in a bowl. Spread the mixture over the chicken and place it in a deep-dish baking pan.
Cover the chicken with aluminum foil and cook in the preheated oven at 330F for 2 hours.
Serve the chicken warm and fresh.
Nutrition: Calories:183, Fat:14.4g, Protein:10.7g, Carbohydrates:3.1g

Tomato Roasted Feta

Preparation time:5 minutes
Cooking time:45 minutes
Servings: 4
Ingredients:
8 oz. feta cheese
2 tomatoes, peeled and diced
2 garlic cloves, chopped
1 cup tomato juice
1 thyme sprig
1 oregano sprig
Directions:
Mix the tomatoes, garlic, tomato juice, thyme and oregano in a small deep-dish baking pan.
Place the feta in the pan as well and cover with aluminum foil.
Cook in the preheated oven at 350F for 10 minutes.
Serve the feta and the sauce fresh.
 Nutrition: Calories:173, Fat:12.2g, Protein:9.2g, Carbohydrates:7.8g

Feta Stuffed Pork Chops

Preparation time:5 minutes
Cooking time:1hour
Serves:4
Ingredients:
4 pork chops
1 teaspoon dried oregano
1 teaspoon dried basil
2 tablespoons chopped dill
1 tablespoon chopped parsley
4 oz. feta cheese, crumbled
1 pinch chili flakes
Directions:
Season the pork chops with oregano and basil then cut a small pocket into each of them.
Mix the dill, parsley, feta and chili in a bowl.

Stuff the pork chops with the feta mixture.
Heat a grill pan over medium flame and place the pork chops on the grill.
Cook on each side for 7-8 minutes.
Serve the pork chops warm with your favorite side dish.
Nutrition: Calories:336, Fat:26.0g, Protein:22.4g, Carbohydrates:2.3g

Creamy Smoked Salmon Pasta

Preparation time:5 minutes
Cooking time:30 minutes
Servings: 4
Ingredients:
2 tablespoons olive oil
2 garlic cloves, chopped
1 shallot, chopped
4 oz. smoked salmon, chopped
1 cup green peas
1 cup heavy cream
Salt and pepper to taste
1 pinch chili flakes
8 oz. penne
Directions:
Heat the oil in a skillet and add the garlic and shallot. Cook for 5 minutes until softened.
Add the salmon and peas, as well as salt and chili flakes.
Cook for 5 more minutes then add the cream.
Lower the heat and cook for 5 more minutes.
In the meantime, cook the penne in a large pot of water just until al dente.
Drain well then mix the pasta with the salmon sauce.
Serve the pasta fresh.
Nutrition: Calories:393, Fat:20.8g, Protein:14.3g, Carbohydrates:38.0g

Potato Salmon Casserole

Preparation time:5 minutes
Cooking time:55 minutes
Servings: 8
Ingredients:
2 tablespoons olive oil
4 potatoes, peeled and sliced
4 salmon fillets, cubed
½ teaspoon chili flakes
½ teaspoon cumin powder
1 teaspoon dried oregano
1 teaspoon dried basil
1 cup heavy cream
1 cup Greek yogurt
2 eggs
Salt and pepper to taste
Directions:
Drizzle the oil in a deep-dish baking pan.
Layer the potatoes and salmon in the pan, sprinkling chili flakes, cumin, oregano and basil between the layers.
Mix the cream, yogurt and eggs in a bowl. Add salt and pepper then pour this mixture over the potatoes and salmon.
Cook in the preheated oven at 350F for 35 minutes.
Serve the casserole warm and fresh.
 Nutrition: Calories:309, Fat:16.3g, Protein:23.3g, Carbohydrates:18.4g

Mediterranean Roasted Lamb and Sweet Potatoes

Preparation time:5 minutes
Cooking time:2 hours
Servings: 8
Ingredients:
2 pounds lamb shoulder
4 sweet potatoes, peeled and cubed
6 garlic cloves, crushed
1 rosemary sprig

1 sage sprig
3 tablespoons olive oil
¼ cup vegetable stock
Salt and pepper to taste
1 teaspoon smoked paprika
½ teaspoon chili powder
Directions:
Season the lamb with salt, pepper, paprika and chili powder.
Combine the potatoes, garlic, rosemary and sage in a deep-dish baking pan.
Place the lamb over the potatoes and add the rosemary and sage.
Cover the pan with aluminum foil and cook in the preheated oven at 330F for 1 ½ hours.
Serve the lamb and potatoes fresh.
Nutrition: Calories:350, Fat:13.8g, Protein:33.2g, Carbohydrates:22.1g

Roasted Chicken in Salt Crust

Preparation time:5 minutes
Cooking time:2 ½ hours
Servings: 8
Ingredients:
1 whole chicken
2 cups salt
1 cup all-purpose flour
4 eggs
½ cup water
2 tablespoons chopped tarragon
1 teaspoon dried thyme
1 teaspoon dried basil
Directions:
Mix the salt, flour, eggs, water, tarragon, thyme and basil in a bowl.
Place a layer of this dough in a deep-dish baking tray.
Arrange the chicken over the chicken and cover with the remaining salt mixture.
Cook in the preheated oven at 330F for 2 hours.

Serve the chicken warm and fresh.
Nutrition: Calories:123, Fat:3.7g, Protein:9.5g, Carbohydrates:12.4g

Herbed Buttery Chicken Legs

Preparation time:5 minutes
Cooking time:1 hour
Servings: 8
Ingredients:
8 chicken legs
½ cup butter, softened
2 tablespoons olive oil
6 garlic cloves, minced
1 teaspoon dried thyme
1 tablespoon chopped cilantro
2 tablespoons chopped parsley
2 tablespoons chopped dill
Salt and pepper to taste
Directions:
Mix the butter, oil, garlic, thyme, cilantro, parsley, dill, salt and pepper in a bowl.
Clean and wash the chicken legs then place them on a chopping board.
Carefully lift the skin of each leg and stuff the cavity with a bit of herbed butter.
Place the chicken in a deep-dish baking pan.
Cook in the preheated oven at 330F for 45 minutes.
Serve the chicken legs warm and fresh.
Nutrition: Calories:328, Fat:22.5g, Protein:29.4g, Carbohydrates:1.3g

Roasted Red Bell Pepper Chicken Stew

Preparation time:5 minutes
Cooking time:1 hour
Servings: 8
Ingredients:

4 chicken breasts, halved
4 tablespoons olive oil
1 sweet onion, chopped
4 garlic cloves, minced
1 jalapeno pepper, chopped
1 can diced tomatoes
1 jar roasted red bell peppers, sliced
1 bay leaf
1 thyme sprig
¼ cup dry white wine
1 ½ cups chicken stock
1 tablespoon tomato paste
Salt and pepper to taste
Directions:
Heat the oil in a skillet and stir in the chicken. Cook on all sides until golden brown then add the onion and garlic. Cook for another 5 minutes.
Add the remaining ingredients and season with salt and pepper. Cook on low heat for 40 minutes.
Serve the stew warm and fresh.
Nutrition: Calories:175, Fat:10.9g, Protein:15.1g, Carbohydrates:3.1g

Rosemary Roasted New Potatoes

Preparation time:5 minutes
Cooking time:1 hour
Servings: 6
Ingredients:
2 pounds new potatoes, washed
3 tablespoons olive oil
2 rosemary sprigs
4 garlic cloves, crushed
Salt and pepper to taste
Directions:
Place the new potatoes in a large pot and cover them with water. Cook for 15 minutes then drain well.
Heat the oil in a skillet and add the rosemary and garlic.

Stir in the potatoes and continue cooking on medium flame for 20 minutes or until evenly golden brown.
Serve the potatoes warm.
Nutrition: Calories:168, Fat:7.2g, Protein:2.7g, Carbohydrates:24.6g

Grilled Pesto Salmon

Preparation time:5 minutes
Cooking time:30 minutes
Servings: 4
Ingredients:
4 salmon fillets
4 tablespoons pesto sauce
Salt and pepper to taste
Directions:
Season the salmon with salt and pepper and spread the pesto over the fish.
Heat a grill pan over medium flame and place the salmon on the grill.
Cook on each side for 5 minutes.
Serve the salmon fresh and warm.
Nutrition: Calories:303, Fat:17.5g, Protein:36.0g, Carbohydrates:1.0g

Slow Cooked Cod Stew

Preparation time:5 minutes
Cooking time:1 hour
Servings: 8
Ingredients:
3 tablespoons olive oil
1 shallot, chopped
2 garlic cloves, minced
2 carrots, sliced
2 celery stalks, sliced
4 tomatoes, sliced
1 teaspoon Worcestershire sauce

1 cup vegetable stock
Salt and pepper to taste
1 bay leaf
1 thyme sprig
8 cod fillets
Directions:
Heat the oil in a skillet and add the shallot and garlic. Cook for 2 minutes until softened then add the carrots, celery, tomatoes, sauce and stock, as well as salt and pepper.
Add the bay leaf and thyme sprig and bring it to a boil.
Cook for 5 minutes then place the fish on top.
Cover with aluminum foil and cook in the preheated oven at 300F for 40 minutes.
Serve the stew warm and fresh.
Nutrition: Calories:255, Fat:6.9g, Protein:41.9g, Carbohydrates:4.7g

Mediterranean Tuna Steaks

Preparation time:5 minutes
Cooking time:20 minutes
Servings: 2
Ingredients:
2 tuna steaks
1 teaspoon dried tarragon
1 teaspoon dried basil
Salt and pepper to taste
2 tablespoons olive oil
Directions:
Season the tuna with salt, pepper, tarragon and basil then drizzle with olive oil.
Heat a grill pan over medium flame then place the tuna steaks on the grill and cook on each side for 2 minutes.
Serve the tuna fresh.
Nutrition: Calories:268, Fat:19.0g, Protein:24.0g, Carbohydrates:0.2g

Herbed Chicken Stew

Preparation time:5 minutes
Cooking time:1 hour
Servings: 6
Ingredients:
3 tablespoons olive oil
6 chicken legs
2 shallots, chopped
4 garlic cloves, minced
2 tablespoons pesto sauce
½ cup chopped cilantro
½ cup chopped parsley
2 tablespoons lemon juice
4 tablespoons vegetable stock
Salt and pepper to taste
Directions:
Heat the oil in a skillet and place the chicken in the hot oil. Cook on each side until golden brown then add the shallots, garlic and pesto sauce.
Cook for 2 more minutes then add the rest of the ingredients. Season with salt and pepper and continue cooking on low heat, covered with a lid, for 30 minutes.
Serve the stew warm and fresh.
 Nutrition: Calories:357, Fat:19.6g, Protein:41.4gCarbohydrates:2.0g

Pesto Braised Roasted Leg

Preparation time:5 minutes
Cooking time:2 ½ hour
Servings: 8
Ingredients:
3 pounds lamb shoulder
6 garlic cloves
1 cup fresh basil
2 tablespoons pine nuts
¼ cup olive oil

1 lemon, juiced
Salt and pepper to taste
Directions:
Mix the garlic, basil, pine nuts, oil and lemon juice in a blender. Pulse until well mixed.
Spread the mixture over the lamb and season with salt and pepper.
Cover the lamb with aluminum foil and cook in the preheated oven at 300F for 2 hours.
Serve the lamb fresh and warm with your favorite side dish.
Nutrition: Calories:389, Fat:20.3g, Protein:48.3g, Carbohydrates:1.1g

Pear Braised Pork

Preparation time:5 minutes
Cooking time:2 ½ hours
Servings: 10
Ingredients:
3 pounds pork shoulder
4 pears, peeled and sliced
2 shallots, sliced
4 garlic cloves, minced
1 bay leaf
1 thyme sprig
½ cup apple cider
Salt and pepper to taste
Directions:
Season the pork with salt and pepper.
Combine the pears, shallots, garlic, bay leaf, thyme and apple cider in a deep-dish baking pan.
Place the pork over the pears then cover the pan with aluminum foil.
Cook in the preheated oven at 330F for 2 hours.
Serve the pork and the sauce fresh.
Nutrition: Calories:455, Fat:29.3g, Protein:32.1g, Carbohydrates:14.9g

Saucy Kidney Beans with Kale

Preparation Time: 20 minutes
Servings: 5)
Nutrition: 239 Calories; 7.3g Fat; 34.6g Carbs; 12.3g Protein; 9g Sugars; 9.2g Fiber
Ingredients
2 tablespoons olive oil
1/2 teaspoon cumin seeds
1/2 teaspoon Heirloom pepper seeds
1 teaspoon garlic, pressed
1/2 cup shallots, finely chopped
2 vine-ripe tomatoes, pureed
1 tablespoon brown sugar
1/2 teaspoon smoked paprika
Sea salt and ground black pepper, to taste
1 Turkish bay laurel
20 ounces kidney beans, rinsed and drained
10 ounces fresh kale leaves
Directions
Heat a large-sized saucepan over medium-high heat; add the oil and swirl it around the saucepan. Then, sauté the cumin seeds, Heirloom pepper seeds, garlic, and shallots until they are aromatic.
Then, add the tomatoes, brown sugar, paprika, salt, black pepper, and bay laurel to the saucepan; bring to a boil. Immediately reduce the heat; let it simmer approximately 7 minutes.
Stir in the kidney beans and kale; let it cook, covered, until the kale leaves have turned a vibrant green color. Enjoy!

Black-Eyed Pea Bowl with Scallions

Preparation Time: 30 minutes+ chilling time
Servings: 6

Nutrition: 370 Calories; 11.7g Fat; 50g Carbs; 18.3g Protein; 7.2g Sugars; 19g Fiber
Ingredients
1-pound black-eyed peas
1 bunch of scallions, sliced
2 garlic cloves, pressed
2 tablespoons fresh cilantro, chopped
1/4 cup ripe olives, pitted and chopped
1/2 teaspoon mixed peppercorns, crushed
1 teaspoon chili powder
Sea salt, to taste
1/4 cup extra-virgin olive oil
4 tablespoons red wine vinegar
1/2 cup plain Greek yogurt
Directions
Place the black-eyed peas and 10 cups of water in a large-sized stockpot over medium-high heat; bring to a boil. Immediately reduce the heat to the lowest setting. Cook for 20 to 25 minutes; drain.
Add the other ingredients, except for the Greek yogurt. Toss until everything is well incorporated.
Top with Greek yogurt; serve well-chilled or keep in your refrigerator until ready to serve. Bon appétit!

Barley with Perfect Roasted Vegetables

Preparation Time: 1 hour 15 minutes
Servings: 4
Nutrition: 236 Calories; 15.5g Fat; 47g Carbs; 8.3g Protein; 10.3g Sugars; 12.1g Fiber
Ingredients
1 eggplant, sliced
2 cups button mushrooms, halved

1 cup grape tomatoes
2 bell peppers, seeded and sliced
1/2 cup green onions, whole
4 garlic cloves
Sea salt and ground black pepper, to taste
1 tablespoon fresh sage, chopped
1 teaspoon red chili pepper flakes
4 tablespoons extra-virgin olive oil
3/4 cup pot barley, soaked and rinsed
1 tablespoon balsamic vinegar
1 tablespoon freshly squeezed lemon juice
2 tablespoons fresh cilantro, chopped
4 tablespoons olives, pitted and sliced
Directions
Arrange your vegetables in a baking pan. Toss them with black pepper, salt, sage, red chili pepper flakes, and 2 tablespoons of olive oil. Roast in the preheated oven at 420 degrees F for 25 to 28 minutes.
In the meantime, bring the pot barley and 3 cups of water to a rapid boil; immediately reduce the heat and let it simmer approximately 45 minutes. Drain and reserve.
Place the cooked barley in a serving bowl. Top with the roasted veggies; drizzle the remaining 2 tablespoons of olive oil, balsamic vinegar and lemon juice over your vegetables.
Lastly, garnish with fresh cilantro leaves and sliced olives. Bon appétit!

Savory Oatmeal with Mozzarella Cheese

Preparation Time: 15 minutes
Servings: 4
Nutrition: 237 Calories; 13.3g Fat; 30.6g Carbs; 13.3g Protein; 3g Sugars; 7.2g Fiber
Ingredients
1 cup rolled oats
2 cups water
2 tomatoes, diced
2 scallion stalks, chopped

1/2 teaspoon red pepper flakes
1/4 teaspoon fennel seeds
1/4 teaspoon salt
2 sprigs thyme, crushed
2 cups mustard greens, torn into pieces
1/2 cup black olives, pitted and sliced
1 tablespoon fresh parsley leaves, chopped
1 tablespoon wine vinegar
2 tablespoons extra-virgin olive oil
1/2 cup mozzarella cheese
Directions
Bring 2 cups of water to a boil over medium-high heat. Then, turn the heat down to low and cook the rolled oats until they thicken, for 4 to 5 minutes.
Stir in the other ingredients, except for the mozzarella cheese; let it cook for 5 minutes longer.
Spoon your oats into serving bowls. Top each serving with mozzarella cheese. Serve warm.

Couscous Casserole with Peppers and Goat Cheese

Preparation Time: 1 hour 25 minutes
Servings: 5
Nutrition: 581 Calories; 14.8g Fat; 86g Carbs; 26.3g Protein; 4.3g Sugars; 11.5g Fiber
Ingredients
1/2-pound black beans, soaked overnight and rinsed
2 tablespoons olive oil
1 cup Vidalia onions, peeled and sliced
1/4 cup red wine
4 tablespoons tomato sauce
1 teaspoon ginger garlic paste
2 cups couscous
1 cup roasted sweet mini peppers, seeded and thinly sliced
1 roasted chili pepper, seeded and thinly sliced
2 tablespoons fresh parsley, chopped
1 teaspoon rosemary, chopped

1/4 teaspoon ground bay leaf
1/4 teaspoon ajwain seeds
1/4 teaspoon cumin seeds
Salt, to taste
4 ounces goat cheese, crumbled
2 tablespoons fresh chives, chopped
Directions
Add the rinsed black beans to a large-sized stockpot and cover with 4 inches of water. Bring it to a boil; now, turn the heat down to the lowest setting; let it simmer for approximately 1 hour.
In the meantime, heat 1 tablespoon of olive oil in a medium-sized saucepan over moderate flame. Now, sauté the Vidalia onions until tender and translucent. Add a splash of wine to deglaze the pan.
Now, add in the tomato sauce, ginger-garlic paste, couscous, and roasted peppers. Season with the parsley, rosemary, ground bay leaf, ajwain seeds, cumin seeds, and salt. Cook an additional 5 minutes.
Fold in the cooked beans and stir with a spoon. Brush the bottom and sides of a baking dish with the remaining tablespoon of olive oil.
Dump the mixture into the prepared baking dish. Scatter the crumbled goat cheese over the top of your casserole.
Bake in the preheated oven at 380 degrees F for 15 to 18 minutes or until it is bubbly. Garnish with fresh chives. Bon appétit!

Orzo Pilaf with Herbs

Preparation Time: 30 minutes
Servings: 4
Nutrition: 308 Calories; 6.9g Fat; 56g Carbs; 6g Protein; 1.6g Sugars; 3.2g Fiber
Ingredients
1 teaspoon coconut oil
1 ½ cups orzo
1 tablespoon olive oil

1 carrot, shredded
1/2 cup scallions, chopped
1/2 teaspoon garlic, pressed
1/2 cup hot water
1 can (14 ½-ounce) reduced-sodium vegetable broth
1 teaspoon rosemary
1 teaspoon basil
2 tablespoons fresh chives, roughly chopped
2 tablespoons fresh cilantro, roughly chopped
Directions
In a large-sized skillet, warm the coconut oil until sizzling. Now, sauté the orzo for 2 to 3 minutes.
Add in the olive oil; once hot, cook the carrot and scallions for 3 to 4 minutes, stirring periodically. After that, add the garlic and cook an additional 40 seconds, stirring frequently to avoid burning the garlic.
Pour the hot water and vegetable broth into the skillet; bring the mixture to a rapid boil.
Stir in the rosemary and basil. Reduce the temperature to low; simmer, covered, for 18 to 22 minutes or until the cooking liquid is absorbed.
Serve garnished with fresh chives and cilantro. Enjoy!

Vermicelli with Beans and Lemon Crema Fresca

Preparation Time: 20 minutes
Servings: 5
Nutrition: 461 Calories; 6.1g Fat; 95g Carbs; 6.8g Protein; 19g Sugars; 10.2g Fiber
Ingredients
3 teaspoons olive oil
1 zucchini, sliced
1 teaspoon basil
1 teaspoon oregano
1 dried laurel
1/2 teaspoon Nora pepper, crushed
2 cloves garlic, minced

16 ounces vermicelli
1 cup boiling vegetable stock
1 (14.5-ounce) can stewed tomatoes
1 (13-ounces) can pinto beans, drained and rinsed
1/2 cup Creme Fraiche
1 teaspoon freshly squeezed lemon juice
1/2 teaspoon finely grated lemon zest
1 tablespoon fresh parsley leaves, chopped
Directions
In a deep saucepan, warm 2 teaspoons of the olive oil; once hot, cook the zucchini until it is softened, 5 to 6 minutes; reserve, keeping warm.
In the same saucepan, heat the remaining 1 teaspoon of olive oil; cook the herbs and garlic until fragrant, stirring frequently; it will take 30 to 40 seconds.
Add the vermicelli, boiling vegetable stock, and stewed tomatoes; put a lid on the saucepan, leaving a small gap.
Now, reduce the heat to medium-low heat, giving it a couple of stirs around. Cook until the vermicelli is al dente or about 7 minutes.
After that, fold in the rinsed beans along with the reserved zucchini; continue simmering until everything is heated through.
In the meantime, mix the Creme Fraiche, lemon juice, and lemon zest; serve the hot vermicelli with the Lemon Crema Fresca on the side; garnish with parsley and enjoy!

Cornbread Squares with Vegetables

Preparation Time: 1 hour 20 minutes
Servings: 6
Nutrition: 260 Calories; 8.1g Fat; 40.5g Carbs; .5g Protein; 4.2g Sugars; 4.4g Fiber
Ingredients
1 cup milk
3 cups boiling vegetable stock
2 cups cornmeal

1 tablespoon ghee
1 tablespoon olive oil
2 cups zucchini, sliced
2 sweet Italian peppers, diced
A bunch of green onions, chopped
2 green garlic stalks, chopped
2 ripe tomatoes, pureed
1 teaspoon basil
1 teaspoon rosemary
Directions
Bring the milk and vegetable stock to a boil in a saucepan.
Slowly and gradually whisk in the cornmeal; turn the heat to the lowest setting and cook approximately 15 minutes or until it has thickened.
Add in the ghee and cook for a further 2 minutes or until heated through. Pour your cornmeal into a greased baking pan and smooth out evenly. Place in your refrigerator for 1 hour or until firm.
Meanwhile, warm the olive oil in a cast-iron skillet over moderate heat. Once hot, cook the zucchini, peppers, onion, and green garlic until they have softened.
Fold in the pureed tomatoes, basil, and rosemary, and continue to cook an additional 2 minutes.
Cut the chilled polenta into squares. Spoon the vegetable mixture over the polenta squares and serve. Bon appétit!

Turkish Pilaf with Roasted Chickpeas

Preparation Time: 20 minutes
Servings: 4
Nutrition: 398 Calories; 8.2g Fat; 72.1g Carbs; 7.2g Protein; 1.7g Sugars; 3.6g Fiber
Ingredients
1 tablespoon ghee, softened
1 ½ cups Baldo rice
4 tablespoons şehriye
3 cups hot water

2 vegetable bouillon cubes
1 tablespoon olive oil
1/2 cup shallots, chopped
2 cloves garlic, minced
1 teaspoon paprika
1/4 teaspoon sumac
1 teaspoon dried mint
1/4 teaspoon coarsely ground cumin seed
Sea salt and red pepper flakes, to taste
1/4 cup roasted chickpeas, roughly chopped

Directions

Melt the ghee in a large-sized skillet over a moderate flame, Now, cook the Baldo rice and şehriye, stirring continuously, until they turn a nice golden color.

Pour 3 cups of hot water into the skillet; fold in the vegetable bouillon cubes. Allow the Baldo rice and şehriye to simmer on the lowest setting until all the water is absorbed by the rice; let it sit for 5 to 10 minutes; fluff the Baldo rice and reserve.

Then, in the same skillet, heat the olive oil until sizzling; cook the shallots until tender and fragrant or about 3 minutes.

Then, add the garlic and the remaining seasonings; let it cook for a further 40 seconds or until everything is aromatic.

Top with the roasted chickpeas and eat immediately. Enjoy!

Chapter 8. Dinner

Grilled Vegetable Feta Tart

Preparation time: 5 minutes
Cooking time: 1 ½ hours
Servings: 8
Ingredients:
Crust:
2 cups all-purpose flour
1 teaspoon instant yeast
½ teaspoon salt
1 cup water
2 tablespoons olive oil
Topping:
1 zucchini, sliced
2 tomatoes, sliced
1 shallot, sliced
1 teaspoon dried basil
1 teaspoon dried oregano
2 garlic cloves, minced
2 tablespoons tomato paste
4 oz. feta cheese, crumbled
Directions:
For the crust, combine all the ingredients in a bowl and mix well. Knead for a few minutes until elastic.
Allow the dough to rest and rise for 20 minutes then roll it into a thin round of dough.
Place the dough on a baking tray.
Mix the garlic, basil, oregano and tomato paste in a bowl. Spread the mixture over the dough.
Heat a grill pan over medium flame and place the zucchini and tomatoes on the grill. Cook for a few minutes on all sides until browned.
Top the tart with the vegetables and shallot then sprinkle the cheese on top.
Bake in the preheated oven at 350F for 25 minutes.
Serve the tart warm and fresh.

Nutrition: Calories:198, Fat:7.0g, Protein:6.3g, Carbohydrates:28.0g

Yogurt Baked Eggplants

Preparation time:5 minutes
Cooking time:45 minutes
Servings: 4
Ingredients:
2 eggplants
4 garlic cloves, minced
1 teaspoon dried basil
2 tablespoons lemon juice
Salt and pepper to taste
1 cup Greek yogurt
2 tablespoons chopped parsley
Directions:
Cut the eggplants in half and score the halves with a sharp knife. Season the eggplants with salt and pepper, as well as the basil then drizzle with lemon juice and place the eggplant halves on a baking tray.
Spread the garlic over the eggplants and bake in the preheated oven at 350F for 20 minutes.
When done, place the eggplants on serving plates and top with yogurt and parsley.
Serve the eggplants right away.
Nutrition: Calories:113, Fat:1.6g, Protein:8.1g, Carbohydrates:19.4g

Asparagus Baked Plaice

Preparation time:5 minutes
Cooking time:45 minutes
Servings: 4
Ingredients:
4 plaice fillets
2 cups cherry tomatoes
1 bunch asparagus, trimmed and halved

½ lemon, juiced
2 tablespoons olive oil
Salt and pepper to taste
Directions:
Combine the tomatoes, asparagus, lemon juice and oil in a deep-dish baking pan. Season with salt and pepper.
Place the fillets on top and cook in the preheated oven at 350F for 15 minutes.
Serve the plaice and the veggies warm and fresh.
Nutrition: Calories:113, Fat:1.6g, Protein:8.1g, Carbohydrates:19.4g

Vegetable Turkey Casserole

Preparation time:5 minutes
Cooking time:1 ½ hours
Servings: 8
Ingredients:
3 tablespoons olive oil
2 pounds turkey breasts, cubed
1 sweet onion, chopped
3 carrots, sliced
2 celery stalks, sliced
2 garlic cloves, chopped
½ teaspoon cumin powder
½ teaspoon dried thyme
2 cans diced tomatoes
1 cup chicken stock
1 bay leaf
Salt and pepper to taste
Directions:
1. Heat the oil in a deep heavy pot and stir in the turkey.
2. Cook for 5 minutes until golden on all sides then add the onion, carrot, celery and garlic. Cook for 5 more minutes then add the rest of the ingredients.
3. Season with salt and pepper and cook in the preheated oven at 350F for 40 minutes.
4. Serve the casserole warm and fresh.

Nutrition: Calories:186, Fat:7.3g, Protein:20.1g, Carbohydrates:9.9g

Mushroom Pilaf

Preparation time:5 minutes
Cooking time:50 minutes
Servings: 4
Ingredients:
2 tablespoons olive oil
1 shallot, chopped
2 garlic cloves, minced
1-pound button mushrooms
1 cup brown rice
2 cups chicken stock
1 bay leaf
1 thyme sprig
Salt and pepper to taste
Directions:
Heat the oil in a skillet and stir in the shallot and garlic. Cook for 2 minutes until softened and fragrant.
Add the mushrooms and rice and cook for 5 minutes.
Add the stock, bay leaf and thyme, as well as salt and pepper and continue cooking for 20 more minutes on low heat.
Serve the pilaf warm and fresh.
Nutrition: Calories:265, Fat:8.9g, Protein:7.6g, Carbohydrates:41.2g

Fig and Prosciutto Pita Bread Pizza

Preparation time:5 minutes
Cooking time:20 minutes
Servings: 6
Ingredients:
4 pita breads
8 figs, quartered
8 slices prosciutto
8 oz. mozzarella, crumbled

Directions:
Place the pita breads on a baking tray.
Top with crumbled cheese then figs and prosciutto.
Bake in the preheated oven at 350F for 8 minutes.
Serve the pizza right away.
Nutrition: Calories:445, Fat:13.7, Protein:39.0g, Carbohydrates:41.5g

Spaghetti in Clam Sauce

Preparation time:5 minutes
Cooking time:45 minutes
Servings: 4
Ingredients:
8 oz. spaghetti
2 tablespoons olive oil
2 garlic cloves, minced
2 tomatoes, peeled and diced
1 cup cherry tomatoes, halved
1-pound fresh clams, cleaned and rinsed
2 tablespoons white wine
1 teaspoon sherry vinegar
Directions:
Heat the oil in a heavy saucepan and add the garlic. Cook for 30 seconds until fragrant then add the tomatoes, wine and vinegar. Bring to a boil and cook for 5 minutes then stir in the clams and continue cooking for 10 more minutes.
In the meantime, bring a large pot of water to a boil with a pinch of salt and add the spaghetti.
Cook them for 8 minutes just until al dente. Drain well and mix with the clam sauce.
Serve the dish right away.
Nutrition: Calories:305, Fat:8.8g, Protein:8.1g, Carbohydrates:48.3g

Creamy Fish Gratin

Preparation time:5 minutes

Cooking time:1 hour
Servings: 6
Ingredients:
1 cup heavy cream
2 salmon fillets, cubed
2 cod fillets, cubed
2 sea bass fillets, cubed
1 celery stalk, sliced
Salt and pepper to taste
½ cup grated Parmesan
½ cup feta cheese, crumbled
Directions:
Combine the cream with the fish fillets and celery in a deep-dish baking pan.
Add salt and pepper to taste then top with the Parmesan and feta cheese.
Cook in the preheated oven at 350F for 20 minutes.
Serve the gratin right away.
Nutrition: Calories:301, Fat:16.1g, Protein:36.9g, Carbohydrates:1.3g

Broccoli Pesto Spaghetti

Preparation time:5 minutes
Cooking time:35 minutes
Servings: 4
Ingredients:
8 oz. spaghetti
1-pound broccoli, cut into florets
2 tablespoons olive oil
4 garlic cloves, chopped
4 basil leaves
2 tablespoons blanched almonds
1 lemon, juiced
Salt and pepper to taste
Directions:
For the pesto, combine the broccoli, oil, garlic, basil, lemon juice and almonds in a blender and pulse until well mixed and smooth.

Cook the spaghetti in a large pot of salty water for 8 minutes or until al dente. Drain well.
Mix the warm spaghetti with the broccoli pesto and serve right away.
Nutrition: Calories:284, Fat:10.2g, Protein:10.4g, Carbohydrates:40.2g

Spaghetti all'Olio

Preparation time:5 minutes
Cooking time:30 minutes
Servings: 4
Ingredients:
8 oz. spaghetti
3 tablespoons olive oil
4 garlic cloves, minced
2 red peppers, sliced
1 tablespoon lemon juice
Salt and pepper to taste
½ cup grated parmesan cheese
Directions:
Heat the oil in a skillet and add the garlic. Cook for 30 seconds then stir in the red peppers and cook for 1 more minute on low heat, making sure to only infuse them, not to burn or fry them.
Add the lemon juice and remove off heat.
Cook the spaghetti in a large pot of salty water for 8 minutes or as stated on the package, just until they become al dente.
Drain the spaghetti well and mix them with the garlic and pepper oil.
Serve right away.
Nutrition: Calories:268, Fat:11.9g, Protein:7.1g, Carbohydrates:34.1g

Quick Tomato Spaghetti

Preparation time:5 minutes
Cooking time:15 minutes
Servings: 4

Ingredients:
8 oz. spaghetti
3 tablespoons olive oil
4 garlic cloves, sliced
1 jalapeno, sliced
2 cups cherry tomatoes
Salt and pepper to taste
1 teaspoon balsamic vinegar
½ cup grated Parmesan
Directions:
Heat a large pot of water on medium flame. Add a pinch of salt and bring to a boil then add the pasta.
Cook for 8 minutes or until al dente.
While the pasta cooks, heat the oil in a skillet and add the garlic and jalapeno. Cook for 1 minute then stir in the tomatoes, as well as salt and pepper.
Cook for 5-7 minutes until the tomatoes' skins burst.
Add the vinegar and remove off heat.
Drain the pasta well and mix it with the tomato sauce. Sprinkle with cheese and serve right away.
Nutrition: Calories:298, Fat:13.5g, Protein:9.7g, Carbohydrates:36.0g

Chili Oregano Baked Cheese

Preparation time:5 minutes
Cooking time:35 minutes
Servings: 4
Ingredients:
8 oz. feta cheese
4 oz. mozzarella, crumbled
1 chili pepper, sliced
1 teaspoon dried oregano
2 tablespoons olive oil
Directions:
Place the feta cheese in a small deep-dish baking pan.
Top with the mozzarella then season with pepper slices and oregano.

Cover the pan with aluminum foil and cook in the preheated oven at 350F for 20 minutes.
Serve the cheese right away.
Nutrition: Calories:292, Fat:24.2g, Protein:16.2g, Carbohydrates:3.7g

Crispy Italian Chicken

Preparation time:5 minutes
Cooking time:40 minutes
Servings: 4
Ingredients:
4 chicken legs
1 teaspoon dried basil
1 teaspoon dried oregano
Salt and pepper to taste
3 tablespoons olive oil
1 tablespoon balsamic vinegar
Directions:
Season the chicken with salt, pepper, basil and oregano.
Heat the oil in a skillet and add the chicken in the hot oil.
Cook on each side for 5 minutes until golden then cover the skillet with a weight – another skillet or a very heavy lid is recommended.
Place over medium heat and cook for 10 minutes on one side then flip the chicken repeatedly, cooking for another 10 minutes until crispy.
Serve the chicken right away.
Nutrition: Calories:262, Fat:13.9g, Protein:32.6g, Carbohydrates:0.3g

Sea Bass in a Pocket

Preparation time:5 minutes
Cooking time:40 minutes
Servings: 4
Ingredients:
4 sea bass fillets

4 garlic cloves, sliced
1 celery stalk, sliced
1 zucchini, sliced
1 cup cherry tomatoes, halved
1 shallot, sliced
1 teaspoon dried oregano
Salt and pepper to taste
Directions:
Mix the garlic, celery, zucchini, tomatoes, shallot and oregano in a bowl. Add salt and pepper to taste.
Take 4 sheets of baking paper and arrange them on your working surface.
Spoon the vegetable mixture in the center of each sheet.
Top with a fish fillet then wrap the paper well so it resembles a pocket.
Place the wrapped fish in a baking tray and cook in the preheated oven at 350F for 15 minutes. Serve the fish warm and fresh.
Nutrition: Calories:149, Fat:2.8g, Protein:25.2g, Carbohydrates:5.2g

Chicken and Chorizo Casserole

Preparation time:5 minutes
Cooking time:1 hour
Servings: 6
Ingredients:
6 chicken thighs
4 chorizo links, sliced
2 tablespoons olive oil
1 cup tomato juice
2 tablespoons tomato paste
1 bay leaf
1 teaspoon dried thyme
Salt and pepper to taste
Directions:
Heat the oil in a skillet and add the chicken. Cook on all sides until golden then transfer the chicken in a deep-dish baking pan.
Add the rest of the ingredients and season with salt and pepper.

Cook in the preheated oven at 350F for 25 minutes.
Serve the casserole right away.
Nutrition: Calories:424, Fat:27.5g, Protein:39.1g, Carbohydrates:3.6g

Lamb Stuffed Tomatoes with Herbs

Preparation time:5 minutes
Cooking time:1 hour
Servings: 6
Ingredients:
6 large tomatoes
1-pound ground lamb
¼ cup white rice
2 shallots, chopped
2 garlic cloves, minced
1 tablespoon chopped dill
1 tablespoon chopped parsley
1 tablespoon chopped cilantro
1 teaspoon dried mint
Salt and pepper to taste
1 tablespoon lemon juice
2 tablespoons olive oil
1 cup vegetable stock
Directions:
Mix the lamb, rice, shallots, garlic, dill, parsley, cilantro and mint in a bowl. Add salt and pepper to taste.
Remove the top of each tomato then carefully remove the flesh, leaving the skins intact.
Chop the flesh finely and place it in a deep heavy saucepan. Add the lemon juice, as well as salt and pepper to taste.
Stuff the tomatoes with the lamb mixture and place them all in the pan.
Drizzle with oil then pour in the stock.
Cover with a lid and cook on low heat for 35 minutes.
Serve the tomatoes right away.
Nutrition: Calories:248, Fat:10.7g, Protein:23.7g, Carbohydrates:14.6g

Creamy Spinach with Polenta and Poached Egg

Preparation time: 5 minutes
Cooking time: 40 minutes
Servings: 4
Ingredients:
Creamy spinach:
2 tablespoons olive oil
2 garlic cloves, minced
1 red pepper, chopped
4 cups baby spinach
½ cup heavy cream
1 tablespoon all-purpose flour
Salt and pepper to taste
Polenta:
½ cup polenta flour
1 ½ cups water
1 tablespoon olive oil
Salt and pepper to taste
Poached eggs:
3 cups water
1 tablespoon white wine vinegar
4 eggs
Directions:
For the creamy spinach, heat the oil in a skillet and add the garlic and red pepper. Cook on high heat for 1 minute then add the spinach and continue cooking for 5-7 minutes until the spinach is softened and most of the liquid has evaporated.
Mix the cream with the flour and pour it over the spinach.
Cook for 5 more minutes until thickened and creamy.
Adjust the taste with salt and pepper and remove off heat.
For the polenta, heat the water with salt in a saucepan.
When it starts to boil, stir in the oil and polenta flour.
Cook on low heat for 10 minutes.
For the poached eggs, bring the water, vinegar and a pinch of salt to a boil in a saucepan.

Crack open the eggs and drop them in the boiling liquid, one by one, cooking them for 1-2 minutes just until set, but still soft in the center.

To serve, spoon the polenta on serving plates. Top with creamy spinach and finish with a poached egg.

Nutrition: Calories:231, Fat:20.7g, Protein:7.3g, Carbohydrates:5.7g

Summer Fish Stew

Preparation time:5 minutes
Cooking time:1 hour
Servings: 6
Ingredients:
3 tablespoons olive oil
4 garlic cloves, minced
1 red onion, chopped
1 celery stalk, sliced
2 red bell peppers, cored and diced
2 tablespoons tomato paste
2 cups cherry tomatoes
1 cup vegetable stock
Salt and pepper to taste
4 cod fillets, cubed
4 sea bass fillets, cubed
2 tablespoons all-purpose flour
Directions:
Season the fish with salt and pepper then sprinkle it with flour.
Heat the oil in a skillet then place the fish and cook it on all sides until golden brown. It just must be golden brown, not cooked through just yet.
Remove the fish on a platter.
Add the garlic, onion and celery in the same skillet as the fish was in and cook for 2 minutes until fragrant.
Stir in the remaining ingredients and season with salt and pepper. Cook for 10 minutes on low heat then add the fish and cook for another 10 minutes.
Serve the stew warm and fresh.

Nutrition: Calories:318, Fat:10.1g, Protein:45.1g, Carbohydrates:10.3g

Chorizo White Bean Stew

Preparation time:5 minutes
Cooking time:1 hour
Servings: 8
Ingredients:
3 tablespoons olive oil
4 chorizo links, sliced
2 sweet onions, chopped
4 garlic cloves, minced
2 celery stalks, sliced
2 carrots, sliced
2 red bell peppers, cored and diced
2 tablespoons tomato paste
1 can diced tomatoes
2 cans white beans, drained
1 bay leaf
1 teaspoon sherry vinegar
½ teaspoon dried oregano
1 cup chicken stock
Salt and pepper to taste
Directions:
Heat the oil in a deep saucepan and stir in the chorizo. Cook for 5 minutes then add the onions and garlic, as well as celery and carrots.
Cook for another 10 minutes to soften.
Add the rest of the ingredients then season with salt and pepper to taste.
Cook on low heat for 35-40 minutes.
Serve the stew right away or freeze it into individual portions for later serving.
Nutrition: Calories:386, Fat:17.4g, Protein:20.2g, Carbohydrates:38.8g

Roasted Eggplant Red Pepper Penne

Preparation time:5 minutes
Cooking time:40 minutes
Servings: 4
Ingredients:
8 oz. penne
2 eggplants
4 roasted red bell peppers, sliced
½ teaspoon dried oregano
2 tablespoons olive oil
Salt and pepper to taste
Directions:
Heat a large pot of water on medium flame. Add a pinch of salt and bring it to a boil.
Add the penne and cook them until al dente, not more than 8 minutes.
Cut the eggplants in half, season them with salt and pepper and place them in a baking tray.
Cook in the preheated oven at 400F for 15 minutes.
When done, scoop out the flesh and chop it into fine bits. Mix with the sliced bell peppers, oregano and oil then adjust the taste with salt and pepper.
Stir in the cooked penne and serve the pasta right away.
 Nutrition: Calories:292, Fat:8.8g, Protein:9.1g, Carbohydrates:47.3g

Italian Chicken Butternut Pot

Preparation time:5 minutes
Cooking time:1 hour
Servings: 8
Ingredients:
4 chicken breasts, cubed
1 tablespoon all-purpose flour
3 tablespoons olive oil
4 cups butternut squash cubes
1 thyme sprig

1 rosemary sprig
2 garlic cloves, minced
1 teaspoon dried oregano
1 tablespoon balsamic vinegar
1 cup vegetable stock
Salt and pepper to taste
Directions:
Season the chicken with salt and pepper then sprinkle it with flour.
Heat the oil in a pot that can go in the oven then add the chicken. Cook on all sides for 10 minutes then add the rest of the ingredients.
Season with salt and pepper and cover with a lid.
Cook in the preheated oven at 350F for 35 minutes.
Serve the dish warm and fresh.
Nutrition: Calories: 178, Fat:9.1g, Protein:15.4g, Carbohydrates:9.4g

Sticky Skillet Chicken

Preparation time:5 minutes
Cooking time:1 hour
Servings: 4
Ingredients:
4 chicken legs
Salt and pepper to taste
3 tablespoons olive oil
2 garlic cloves, chopped
2 tablespoons honey
2 tablespoons lemon juice
1 thyme sprig
1 rosemary sprig
Directions:
Season the chicken with salt and pepper.
Heat the oil in a skillet and place the chicken in the hot oil.
Fry on each side for 10-15 minutes until golden brown.
Drizzle in the honey and lemon juice then place the herb sprigs on top.

Cover with a lid or aluminum foil and place in the preheated oven at 350F for 20 minutes.
Serve the chicken and the sauce right away.
Nutrition: Calories:316, Fat:18.0g, Protein:29.1g, Carbohydrates:9.3g

Chapter 9. Dietary desserts
Compote Dipped Berries Mix

Preparation Time: 20 minutes
Servings: 8
Ingredients:
2 cups fresh strawberries, hulled and halved lengthwise
4 sprigs fresh mint
2 cups fresh blackberries
1 cup pomegranate juice
2 teaspoons vanilla
6 orange pekoe tea bags
2 cups fresh red raspberries
1 cup water
2 cups fresh golden raspberries
2 cups fresh sweet cherries, pitted and halved
2 cups fresh blueberries
2 ml bottle Sauvignon Blanc
Directions:

Preheat the oven to 290 degrees F and lightly grease a baking dish.
Soak mint sprigs and tea bags in boiled water for about 10 minutes in a covered bowl.
Mix all the berries and cherries in another bowl and keep aside.
Cook wine with pomegranate juice in a saucepan and add strained tea liquid.
Toss in the mixed berries to serve and enjoy.
Nutrition:

Calories 356
Total Fat 0.8 g
Saturated Fat 0.1 g
Cholesterol 0 mg
Total Carbs 89.9 g
Dietary Fiber 9.4 g
Sugar 70.8 g
Protein 2.2 g

Fruity Almond Cake

Preparation Time: 2 hours 10 minutes
Servings: 4
Ingredients:
¾ cup butter, softened
2½ oz. whole almond
1 large orange, zested and juiced
2½ oz. whole wheat flour
1 teaspoon mixed spice
1-pound mixed dried fruit
¾ cup light muscovado sugar
½ vanilla pod, seeds scraped
2 large eggs, beaten
¼ cup sherry
2 oz. ground almond
Directions:
Preheat the oven to 300 degrees F and lightly grease a cake pan with butter.
Mix sherry, fruits, orange juice and zest in a bowl and refrigerate overnight.
Beat sugar and vanilla seeds in butter until creamy and smooth.
Combine mixed spice, whole wheat flour and ground almond until smooth.
Add in the marinated fruits and whole almonds.
Pour the batter in the baking dish and transfer in the oven.
Bake for about 1 hour 30 minutes and dish out.
Reduce the heat of the oven to 270 degrees F and bake for 20 more minutes to enjoy.
Nutrition:
Calories 613
Total Fat 41.8 g
Saturated Fat 19.8 g
Cholesterol 169 mg
Total Carbs 54.1 g
Dietary Fiber 2.7 g
Sugar 36.8 g
Protein 9 g

Orange Sesame Cookies

Preparation Time: 35 minutes
Servings: 12
Ingredients:
½ lemon, juiced1 cup brown sugar
1 cup extra virgin olive oil
½ cup orange juice, freshly squeezed
½ teaspoon ground cloves
½ cup sesame seed
3¾ cups whole wheat flour
½ shot brandy
1 teaspoon baking soda
½ teaspoon ground cinnamon
Directions:
Preheat the oven to 360 degrees F and lightly grease a baking tray.
Beat sugar with olive oil until dissolved and add orange juice.
Beat again for about 2 minutes and stir in lemon juice, cinnamon, cloves, baking soda and brandy.
Fold in the whole wheat flour and mix well to prepare smooth cookie dough.
Make small cookies out of this mixture and roll in the sesame seeds.
Arrange the cookies on the baking tray and transfer in the oven.
Bake for about 25 minutes and dish out to serve and enjoy.
Nutrition:
Calories 375
Total Fat 20.2 g
Saturated Fat 2.9 g
Cholesterol 0 mg
Total Carbs 44.5 g
Dietary Fiber 1.9 g
Sugar 12.8 g
Protein 5.2 g

Popped Quinoa Bars

Preparation Time: 10 minutes
Servings: 3
Ingredients:
2 (4 oz.) semi-sweet chocolate bars, chopped
½ tablespoon peanut butter
½ cup dry quinoa
¼ teaspoon vanilla
Directions:
Toast dry quinoa in a pan until golden and stir in chocolate, vanilla and peanut butter.
Spread this mixture in a baking sheet evenly and refrigerate for about 4 hours.
Break it into small pieces and serve chilled.
Nutrition:
Calories 278
Total Fat 11.8 g
Saturated Fat 6.6 g
Cholesterol 7 mg
Total Carbs 36.2 g
Dietary Fiber 3.1 g
Sugar 15.4 g

Greek Baklava

Preparation Time: 1 hour 10 minutes
Servings: 12
Ingredients:
2 cups walnuts, chopped
1 teaspoon ground cloves
1 cup sesame seeds
3 tablespoons honey
2 cups almonds, chopped
12 sheets phyllo pastry dough
1 cup extra-virgin olive oil
2 teaspoons ground cinnamon Syrup:
2½ cups honey

2 lemons, peeled and juiced
4 cups water
2 cinnamon sticks
Directions:
Preheat the oven to 360 degrees F and lightly grease a baking sheet and phyllo sheets with olive oil.
Mix almonds, cinnamon, walnuts, sesame seeds, cloves and honey in a bowl.
Place this layer in the baking dish and top with 3 more layers of phyllo sheets.
Pour in half of the nut mixture and evenly spread it.
Add layers of 4 oiled phyllo sheets again and top with the other half of the nut mixture.
Transfer in the oven and bake the baklawa for about 35 minutes.
Slice the layers into squares and allow it to cool.
Meanwhile, let simmer all the sauce ingredients for about 15 minutes.
Pour it over the baklava pieces and serve immediately.
Nutrition:
Calories 651
Total Fat 43.8 g
Saturated Fat 4.5 g
Cholesterol 0 mg
Total Carbs 61.3 g
Dietary Fiber 5.6 g
Sugar 40.1 g
Protein 13 g

Honey Yogurt Cheesecake

Preparation Time: 1 hour 10 minutes
Servings: 8
Ingredients:
1 cup Greek yogurt
Fresh fruit, to serve
3 tablespoons almond butter, melted
1 cup honey
3 tablespoons almonds, flaked

2 eggs
2 tablespoons breadcrumbs
1 orange, zested
4 oz. amaretti biscuits
26 oz. mascarpone
1 lemon, zested
Directions:
Preheat the oven to 290 degrees F and lightly grease a baking dish.
Seal biscuits and almonds in a Ziplock bag and crush with a rolling pin.
Toss this mixture with almond butter and breadcrumbs and transfer evenly into a baking dish.
Bake for about 10 minutes and dish out.
Whisk eggs, yogurt and mascarpone with a beater and stir in honey, orange and lemon zest.
Transfer the batter to the baked crust and cover the pan with a foil tent.
Bake for about 1 hour and garnish with honey and almonds to serve.
Nutrition:
Calories 368
Total Fat 30 g
Saturated Fat 11 g
Cholesterol 217 mg
Total Carbs 10 g
Dietary Fiber 8 g
Sugar 0.3 g

Almond Orange Pandoro

Preparation Time: 10 minutes
Servings: 12
Ingredients:
2 large oranges, zested
2½ cups mascarpone
½ cup almonds, whole
2½ cups coconut cream

½ pandoro, diced
2 tablespoons sherry
Directions:
Whisk cream with mascarpone, icing sugar, ¾ zest and half sherry in a bowl.
Dice the pandoro into equal sized horizontal slices.
Place the bottom slice in a plate and top with the remaining sherry.
Spoon the mascarpone mixture over the slice.
Top with almonds and place another pandoro slice over.
Continue adding layers of pandoro slices and cream mixture.
Dish out to serve.
Nutrition:
Calories 346
Total Fat 10.4 g
Saturated Fat 3 g
Cholesterol 10 mg
Total Carbs 8.5 g
Dietary Fiber 3 g
Sugar 2.4 g

Honey Glazed Pears

Preparation Time: 35 minutes
Servings: 3
Ingredients:
2 tablespoons almond butter
3 ripe medium pears, peeled, halved and cored
1 teaspoon orange zest
1/3 cup salted pistachios, roasted and chopped
¼ cup pear nectar
Dollop of cream
3 tablespoons honey
½ cup mascarpone cheese
Directions:
Preheat the oven to 400 degrees F and lightly grease a baking pan.

Spread the sliced pears in a baking pan with their cut sides down.
Top with butter, honey, nectar and orange zest.
Transfer in the oven and roast these pears for about 25 minutes.
Mix sugar with mascarpone and pour on the baked pears.
Garnish with honey and pistachios to serve.
Nutrition:
Calories 349
Total Fat 14.3 g
Saturated Fat 8.4 g
Cholesterol 41 mg
Total Carbs 53.6 g
Dietary Fiber 5.8 g
Sugar 41.5 g
Protein 6 g

Blueberries with Lemon Cream

Servings: 4 servings, 1/2 cup each
Preparation Time: 10 min
Ingredients:
2 cups fresh blueberries
4-ounce cream cheese, s reduced-fat (Neufchâtel)
3/4 cup vanilla yogurt, low-fat
2 teaspoons lemon zest, freshly grated
1 teaspoon honey
Directions:
In a medium mixing bowl, break up the cream cheese with a fork.
Drain the liquid from yogurt. Put the yogurt in a bowl and add the honey. With an electric or hand mixer, beat at HIGH speed until the mixture is creamy and light; Stir the lemon zest.
Layer the blueberries and the lemon cream cheese in wine glasses
Using a fork, break up cream cheese in a medium bowl. Drain off any liquid from the yogurt; add yogurt to the bowl along with

honey. Using an electric mixer, beat at high speed until light and creamy. Stir in lemon zest.
Layer the lemon cream and blueberries in dessert dishes or wineglasses. If not serving immediately, cover and refrigerate for up to 8 hours.
Nutrition: 106 Calories, 3 g total fat (0 g sat. fat, 1 g poly. Fat, 1 g mono), 24 mg Chol., 186 mg sodium, 12 g carb.,7 g fiber,2 g sugar, 15 g protein.

Mediterranean Biscotti

Servings: 3 dozen
Preparation Time: 25 min
Cooking Time: 1 hr.
Ingredients:
2 eggs
1 cup whole-wheat flour
1 cup all-purpose flour
3/4 cup parmesan cheese, grated
2 teaspoons baking powder
2 tablespoons sugar
1/4 cup sun-dried tomato, finely chopped
1/4 cup Kalamata olive, finely chopped
1/3 cup olive oil
1/2 teaspoon salt
1/2 teaspoon black pepper, cracked
1 teaspoon dried oregano (preferably Greek)
1 teaspoon dried basil
Directions:
Into a large-sized bowl, beat the eggs and the sugar together. Pour in the olive; beat until smooth.
In another bowl, combine the flours, baking powder, pepper, salt, oregano, and basil. Stir the flour mix into the egg mixture, stirring until blended.
Stir in the cheese, tomatoes, and olives; stirring until thoroughly combined.

Divide the dough into 2 portions; shape each into 10-inch long logs. Place the logs into a parchment-lined cookie sheet; flatten the log tops slightly.

Bake for about 30 minutes in a preheated 375F oven or until the logs are pale golden and not quite firm to the touch.

Remove from the oven; let cool on the baking sheet for 3 minutes. Transfer the logs into a cutting board; slice each log into 1/2-inch diagonal slices using a serrated knife.

Place the biscotti slices on the baking sheet, return into the 325F oven, and bake for about 20 to 25 minutes until dry and firm. Flip the slices halfway through baking. Remove from the oven, transfer on a wire rack and let cool.

Notes: Store the biscotti slices into airtight containers.

Nutrition: 731.6 Calories, 36.5 g total fat (9 g sat. fat), 146 mg Chol., 1238.4 mg sodium, 77.8 g carb., 3.5 g fiber, 10.7 g sugar, and 23.3 g protein.

Frozen Mediterranean Delight

Servings: 4
Preparation Time: 15 min, plus 4 hours (2 hours for draining, 2 hours for freezing)
Ingredients:
6 pitted dates, chopped
3 cups yogurt, plain, nonfat
2/3 cup pistachios, natural, unsalted, shelled
2 ounces bittersweet chocolate
1/2 cup sugar
1 tablespoon ouzo
Directions:
Line a fine meshed strainer with cheesecloth. Place the strainer over a bowl. Put the yogurt in the cheesecloth lines strainer; allow to drain for 2 hours.

Put the sugar and half of the pistachios in a coffee grinder or a food processor, grind or process until powder.

Roughly chop the remaining pistachios.

Combine the drained yogurt, nuts, sugar-pistachio mixture, ouzo, and dates, mixing until well incorporated; place in the

freezer. After 1 hour, remove from the freezer and mix well. Return to the freezer and freeze until firm.
Divide into 4 servings. Garnish with chocolate shavings; serve.
Nutrition: 349 Calories, 9.5 g total fat (1.3 g sat. fat), 3.7 mg Chol., 141.9 mg sodium, 54.2 g carb., 3.1 g fiber, 48.5 g sugar, and 15.1 g protein.

Chocolate Baklava

Servings: 24(1 piece)
Preparation Time: 46 min
Cooking Time: 35 min
Ingredients:
24 sheets (14 x 9-inch) frozen whole-wheat phyllo (filo) dough, thawed
1/8 teaspoon salt
1/3 cup toasted walnuts, chopped coarsely
1/3 cup almonds, blanched toasted, chopped coarsely
1/2 teaspoon ground cinnamon
1/2 cup water
1/2 cup hazelnuts, toasted, chopped coarsely
1/2 cup pistachios, roasted, chopped coarsely
3/4 cup honey
1/2 cup of butter, melted
1 cup chocolate-hazelnut spread (I used Nutella)
1-piece (3-inch) cinnamon stick
Cooking spray
Directions:
Into medium-sized saucepan, combine the water, honey, and the cinnamon stick; stir until the honey is dissolved. Increase the heat/flame to medium; continue cooking for about 10 minutes without stirring. A candy thermometer should read 230F. Remove the saucepan from the heat and then keep warm.
Remove and discard the cinnamon stick.
Preheat the oven to 350F.
Put the chocolate-hazelnut spread into microwavable bowl; microwave the spread for about 30 seconds on HIGH or until the spread is melted.

In a bowl, combine the hazelnuts, pistachios, almonds, walnuts, ground cinnamon, and the salt.

Lightly grease with the cooking spray a 9x13-inch ceramic or glass baking dish.

Put 1 sheet lengthwise into the bottom of the prepared baking dish, extending the ends of the sheet over the edges of the dish. Lightly brush the sheet with the butter. Repeat the process with 5 sheets phyllo and a light brush of butter. Drizzle 1/3 cup of the melted chocolate-hazelnut spread over the buttered phyllo sheets. Sprinkle about 1/3 of the nut mixture (1/2 cup) over the spread. Repeat the process, layering phyllo sheet, brush of butter, spread, and with nut mixture. For the last, nut mixture top layer, top with 6 phyllo sheets, pressing each phyllo gently into the dish and brushing each sheet with butter.

Slice the layers into 24 portions by making 3 cuts lengthwise and then 5 cuts crosswise with a sharp knife; bake for about 35 minutes at 350F or until the phyllo sheets are golden. Remove the dish from the oven, drizzle the honey sauce over the baklava. Pace the dish on a wire rack and let cool. Cover and store the baklavas at normal room temperature if not serving right away.

Notes: The sheets of phyllo are delicately thin so handle them with care to avoid tearing them. Cover the sheets with damp cloth so they won't dry out while you are working.

Nutrition: 238 Calories, 13.4 g total fat (4.3 g sat. fat, 5.6 g mono fat, 2 g poly fat), 4 g protein, 27.8 g total carbs., 1.6 g fiber, 10 mg Chol., 1.3 mg iron, 148 mg sodium, and 29 mg calcium.

Orange-Glazed Fruit and Ouzo Whipped Cream

Servings: 4
Preparation Time: 20 min, plus 30 min chilling
Cooking Time: 10 min
Ingredients:
3 cups fruit (such as tangerine wedges, quartered apricots or plums, or strips of mango)
1 tablespoon olive oil spread/butter divided (I Can't Believe It's Not Butter! ®), melted
Chopped almonds, optional (or pistachios)

For the ouzo whipped cream:
1 teaspoon sugar
1 teaspoon ouzo liqueur (anise-flavored), orange juice, orange liqueur, or several drops of anise extract
1/2 cup whipping cream
For the sauce:
2 tablespoons sugar
2 tablespoons honey
1/4 cup orange juice
Directions:
For the syrup:
Mix the syrup ingredients inside a small-sized saucepan. Bring the mixture to a boil, stirring, until the honey and the sugar are dissolved and reduce the heat. Simmer the mixture, without cover, for 10 minutes and set aside.
For the ouzo whipped cream:
In a medium-sized chilled bowl, beat the ouzo whipped cream ingredients using electric mixer on medium speed until soft peaks form with the tips curled. Cover and refrigerate for about 30 minutes to chill.
For the grilled fruit:
Toss the melted olive oil butter and the fruit in a mixing bowl. Transfer the fruit into a foil pan (see notes) or grill pan.
If using charcoal grill, put pan with fruits on the uncovered grill rack over medium coals; grill for about 10-12 minutes, stirring occasionally, until the fruits are heated through.
If using gas grill, first, preheat the grill, then reduce to medium heat. Put the grill rack on the grill rack. Cover the grill and grill for about 10-12 minutes, stirring occasionally, until the fruits are heated through.
Divide the fruits between 4 pieces dessert plates and drizzle with the honey syrup. If desired, sprinkle with the almonds. Serve with the ouzo whipped cream.
Notes: I Can't Believe It's Not Butter! ® is a great butter alternative made with oil blends, water, and salt. It's a simple and delicious spread that's all-natural. To make the foil pan, fold a heavy foil into double thickness. Fold the sides up to create a pan and then cut slits in the bottom.

Nutrition: 267 Calories, 15 g total fat (9 g saturated fat, 4 g mono fat, 1 g poly fat), 44 mg sodium, 49 mg Chol., 36 g total carbs., 26 g sugar, 2 g fiber, and 2 g protein.

Lemon Curd Filled Almond-Lemon Cake

Servings: 8(1 wedge)
Preparation Time: 30 min
Cooking Time: 35 min
Ingredients:
4 large egg yolks
4 large egg whites
2 teaspoons matzo cake meal
2 cups fresh raspberries
1/4 teaspoon of salt
1/4 cup matzo cake meal
1/4 cup blanched almonds, ground
1/2 teaspoon grated lemon rind
1 teaspoon lemon juice, fresh
1 cup sugar
1 cup Lemon Curd
1 1/2 teaspoons water
Cooking spray
Directions:
Preheat the oven to 350F.
Coat a 9-inch spring-form pan with the cooking spray. Dust the pan with the 2 teaspoons of matzo cake meal.
Place the yolks into a large-sized bowl; beat with a mixer at high speed for about 2 minutes. Gradually add the sugar and beat the mixture until pale and thick, about 1 minute. Add the 1/4 cup matzo cake meal, water, lemon rind, lemon juice, and salt; beat until the mixture is just blended. Fold in the almonds.
Place the egg whites into a large-sized bowl. With clean, dry beaters, beat the egg whites using a mixer at high speed until stiff peaks form. Gently stir in 1/4 of the egg whites into the yolk mixture; gently fold in the remaining of the egg whites. Spoon the batter into prepared spring-form pan.

Bake for about 35 minutes at 350F or until the cake is set and brown; remove the pan from the oven, place in a wire rack, and let cool for 10 minutes. Run a knife around the edge of the cake, remove the cake from the pan, place in the wire rack and let cool completely. The cake will sink as it cools.

Spread about 1 cup of lemon curd in the center of the cake. Top with the raspberries. Cut the cake into 8 wedges with a serrated knife. Serve immediately.

Notes: You can prepare the curd 1 or 2 days ahead of time. You can enjoy leftovers on fruit or ice cream. You can also bake the cake earlier in the day and let it cool on a wire rack. 7. Decorate the cake with the curd and the berries just before serving.

Nutrition: 238 Calories, 6.6 g total fat (2.1 g sat. fat, 2.7 g mono fat, 1 g poly fat), 5.9 g protein, 41.4 g total carbs., 2.7 g fiber, 149 mg Chol., 1.1 mg iron, 123 mg sodium, and 36 mg calcium.

Greek Almond Rounds Shortbread

Servings: 84
Preparation Time: 45 min, plus 1hr chilling
Cooking Time: 12 min
Ingredients:
1 1/2 cups butter, softened
1 cup blanched almonds, lightly toasted and finely ground
1 cup powdered sugar
2 egg yolks
2 tablespoons brandy or orange juice
2 tablespoons rose flower water, (optional)
2 teaspoons vanilla
3 1/2 cups cake flour
Powdered sugar
Directions:
Using an electric mixer, beat the butter on MEDIUM or HIGH speed for about 30 seconds in a large sized bowl. Add the 1 cup powdered sugar; beat until the mixture is light in color and fluffy, occasionally scraping the bowl as needed.
Beat in the yolks, vanilla, and the brandy until combined.

With a wooden spoon, stir in the flour and almonds until well incorporated. Cover and refrigerate for about 1 hour or until chilled and the dough is easy to handle.
Preheat the oven to 325F.
Shape the dough into 1-inch balls. Place the balls 2 inches apart int an ungreased cookie sheet. Dip a glass in the additional powdered sugar and use it to flatten each ball into 1/4-inch thickness, dipping the bottom of the glass every time you flatten a ball into cookies.
Place the cookie sheet into the preheated oven; bake for about 12-14 minutes or until the cookies are set.
When the cookies are baked, transfer them on wire racks. While they are still warm, brush with the rose water, if desired.
Sprinkle with more powdered sugar. Let cool completely on the wire racks.
Notes: If using rose water, make sure that you use the edible kind. To store, layer the cookies with waxed paper between each cookie and keep on airtight containers. Close the container tightly and store at room temperature for up to 3 days or freeze for up to 3 months.
Nutrition: 62 Calories, 4 g total fat (2.2 g sat. fat), 14 mg Chol., 24 mg sodium, 15 mg pot., 5.7 g total carbs, 1.5 g sugar, and 0.9 g protein.

Tiny Orange Cardamom Cookies

Servings: 80 cookies (5 cookies per serving)
Preparation Time: 48 min
Cooking Time: 12 min
Ingredients:
1/2 cup whole-wheat flour
1/2 cup all-purpose flour
1 large egg
1 tablespoon sesame seeds, toasted, optional (salted roasted pistachios, chopped)
1 teaspoon orange zest
1 teaspoon vanilla extract
1/2 cup butter, softened

1/2 cup sugar
1/4 teaspoon ground cardamom
Directions:
Preheat the oven to 375F.
In a medium bowl, blend the orange zest and the sugar thoroughly, and then blend in the cardamom. Add the butter and with a mixer, beat until the mixture is fluffy and light. Beat in the egg and the vanilla into the mixture. With the mixer on low speed, mix in the flours into the mixture.
Line 3 baking sheets with parchment paper. Using a level teaspoon measure, drop batter of the cookie mixture onto the sheets. Top each cookie with a pinch of sesame seeds or nuts, if desired; bake for 1bout 10-12 minutes or until the cookies are brown at the edges and crisp. When baked, transfer the cookies on a cooling rack and let them cool completely.
Nutrition: 113 Calories, 1.4 g protein, 6.5 g total fat (3.8 g sat. fat) 12 g total carbs., 0.3 g fiber, 46 mg sodium, and 29 mg Chol.

Classic Baklava

Servings: 18
Preparation Time: 30 min
Cooking Time: 50 min
Ingredients:
1/2 cup honey
1 teaspoon vanilla extract
1 teaspoon ground cinnamon
1-pound nuts, chopped
1 package (16 ounce) phyllo dough
1 cup white sugar
1 cup water
1 cup butter
Directions:
Preheat the oven to 350F or 175C.
Butter the sides and the bottom of a 9x13-inch pan.
Toss the nuts and the cinnamon together; set aside.

Unroll the phyllo dough. Cut the whole stack into half to fit into the pan. Cover the phyllo with a damp cloth as you work to keep them from drying.

Place 2 sheets of phyllo into the pan, butter thoroughly. Repeat the process until you have a layer of 8 sheets. Sprinkle about 2 to 3 tablespoons of the nut mixture on top of the phyllo layer. Top the nut layer with 2 phyllo sheets and butter the sheets. Repeat the process until you have about 6 to 8 layers.

With a sharp knife, cut the layers, all the way to the bottom of the pan, into squares; bake for about 50 minutes or until the baklava is crisp and golden.

Meanwhile, boil the water and the sugar until the sugar is melted. Add the honey and the vanilla; simmer for about 20 minutes.

When the baklava is baked, remove from the oven and immediately spoon the sauce over; let cool.

Serve the baklava slices in cupcake papers.

Nutrition: 393 Calories, 25.9 g total fat (9 g sat. fat), 27 mg Chol., 196 mg sodium, 37.5 g total carbs., 19.9 g sugar, 3.1 g fiber, and 6.1 g protein.

Loukoumades

Servings: 25
Preparation Time: 10 min, plus 1 hr. rising
Cooking Time: 15 min
Ingredients:
3 eggs
1 cup warm water (no more than 100F or 40C)
1 teaspoon salt
1/2 cup honey
1/2 cup warm milk
1/2 cup water
1/3 cup butter, softened
1/4 cup white sugar
2 packages (.25 ounce each) active dry yeast
2 teaspoons ground cinnamon
2 cups whole-wheat flour
2 cups all-purpose flour

4 cups vegetable oil, or as needed

Directions:

Pour the warm water into a small-sized bowl. Sprinkle the yeast over the warm water; let stand for 5 minutes or until the yeast is soft and starts to form into a creamy foam.

Into a large-sized bowl, mix the warm milk, salt, and sugar; mixing until the sugar and the salt are dissolved. Pour the yeast mixture into the milk mixture; stir to combine.

Into the yeast-milk mixture, beat in the eggs, butter, and the flours until the mixture is a soft, smooth dough. Cover the bowl; let the dough rise until it has doubled, about 30 minutes. After 30 minutes, stir the dough well, cover again, and let rise for 30 minutes more.

In a saucepan, mix the 1/2 cup water and honey; bring to a boil over medium-high heat. When boiling, turn the heat off and let cool.

Pour enough vegetable oil into a large saucepan or a deep-fryer to make 2-inches deep; heat until the oil is 350F or 175C.

Place a large soup or tablespoon in a glass of water near the batter. Scoop about 2 tablespoons worth of the dough with a wet spoon. Drop the dough into a wet palm of your hand and then roll it back into the spoon, creating a round shape. Make sure not to over handle the dough. In batches, drop the dough balls into the hot oil, wetting the spoon each time you make a ball. Fry the dough balls until the bottoms are golden brown, turn and cook until golden brown, about 2 to 3 minutes per batch. Gently transfer the loukoumades in a paper towel lined plate; set aside. Cook the remaining dough.

When all the loukoumades are fried, transfer them into a baking sheet. Drizzle with the honey syrup and then sprinkle with cinnamon; serve warm.

Nutrition: 167 Calories, 6.9 g total fat (2.3 g sat. fat), 29 mg Chol., 122 mg sodium, 23.5 g total carbs., 7.9 g sugar, 0.8 g fiber, and 3.2 g protein.

Chapter 10. Desserts for special events

Greek Yogurt Frosted Zucchini Cupcakes with

Servings: 12(1 cupcake)
Preparation Time: 20 min
Cooking Time: 20 min
Ingredients:
1 cup quinoa, cooked according to package directions
1 cup whole-wheat flour
1 cup zucchini, shredded
1 teaspoon ground cinnamon
1 teaspoon vanilla
1/2 cup applesauce, unsweetened
1/2 teaspoon baking powder
1/2 teaspoon salt
1/2-1 teaspoon lemon peel, finely shredded
1/4 cup canola oil
1/4 cup fat-free milk
1/4 cup granulated sugar
1/4 cup packed brown sugar
1/4 teaspoon baking soda
2 eggs
Nonstick cooking spray
For the Greek yogurt frosting:
1 carton (6 –ounces) Greek yogurt, plain, fat-free
1 teaspoon vanilla
2 tablespoons light agave nectar or 3 tablespoons powdered sugar
Directions:
For the muffins:
Preheat the oven to 300F.
Light coat 12 pieces 2 1/2-inch muffin tin with the cooking spray; set aside.
In a large-sized bowl, stir the flour, brown sugar, granulated sugar, baking powder, cinnamon, baking soda, and salt; set aside.

In a medium-sized bowl, beat the milk and the eggs together. Add in the zucchini, quinoa, oil, applesauce, and vanilla; stir well until mixed.
Add the zucchini mixture into the flour mixture and gently stir to combine. Spoon the mixture into the prepared muffin cup, filling each cup about 3/4 full.
Place the muffin cups into the oven and bake for about 20 minutes or a toothpick comes out clean when inserted in the centers of the muffins.
When baked, place the muffin cups on a wire rack, let the muffin cool in the cups for 5 minutes. After 5 minutes, loosen the edges of the muffins from the muffin cups, remove them, and let cool completely on the wire rack, about 1 hour. Frost with the frosting and the sprinkle with the lemon peel.
For the frosting:
Whisk together all the Greek yogurt-frosting ingredients in a mixing bowl.
Nutrition: 167 Calories, 6 g total fat (1 g sat. fat), 32 mg Chol., 161 mg sodium, 24 g total carbs., 13 g sugar, 2 g fiber, and 5 g protein.

Apricots and Mascarpone Cream

Servings: 8
Preparation Time: 30 min
Ingredients:
8 fresh apricots, pitted, halved
4 ounces mascarpone cheese
3 tablespoons white sugar
3 tablespoons apricot preserves
1 cup whipping cream
2 tablespoons apricot nectar
1/4 cup honey
1/2 teaspoon vanilla extract
1 tablespoon balsamic vinegar
1 pinch ground cardamom (optional)
Directions:

With an electric or hand mixer, beat the whipping cream inside a chilled bowl until the whipping cream forms soft peaks. Beat the sugar in; set aside.

In a different bowl, with an electric mixer with a clean beater, whip the mascarpone cheese until very soft. Beat in the vanilla extract, apricot nectar, and cardamom. Gently fold in the mascarpone mixture into the whipped cream.

Place the honey and the apricot preserves into a microwavable bowl; heat in a microwave for about 30 seconds or until warm, but not hot. Stir the balsamic vinegar and mix well.

Stuff each apricot half with a dollop of mascarpone cream mixture. Place them on a serving dish and drizzle the fruit and the plate with the balsamic sauce; serve.

Nutrition: 253 Calories, 17.7 g total fat (10.4 g sat. fat), 58 mg Chol., 23 mg sodium, 23.9 g total carbs., 20.7 g sugar, 0.8 g fiber, and 2.2 g protein.

Minty Orange Greek Yogurt

Servings: 1
Preparation Time: 5 min
Ingredients:
6 tablespoons Greek yogurt, fat-free
4 fresh mint leaves, thinly sliced
1 large orange, peeled, quartered, and then sliced crosswise
1 1/2 teaspoons honey
Directions:
Stir together the honey and the yogurt.
Place the orange slices into a dessert glass. Spoon the honeyed yogurt over the orange slices in the glass and scatter the mint on top of the yogurt.
Nutrition: 171 Calories, 34 g total carbs, 5 g fiber, and 11 g protein.

Apricot Almond Dips

Servings: 24 dips
Preparation Time: 15 min, plus 15 min chilling

Ingredients:
1 package (7 ounces) dried and pitted Mediterranean apricots
4 ounces white candy coating
24 whole almonds, toasted
Directions:
Stuff each apricot with 1 piece with almond.
Melt the candy coating in a microwave, stirring until smooth.
Line a baking sheet with wax paper.
Dip each almond-stuffed apricot in the coating, letting the excess coat drip off. Place in the prepared baking sheet.
Refrigerate and chill for 15 minutes or until the coating is set.
Store the dips in a refrigerator.
Nutrition: 53 Calories, 2 g total fat (1 g sat. fat), 0 mg Chol., 7 mg sodium, 9 g total carbs., 7 g sugar, 1 g fiber, and 1 g protein.

Rustic Raspberry and Fig Mini Crostatas

Servings: 10
Preparation Time: 45 min, plus 2 hours refrigerating
Cooking Time: 30-35 min
Ingredients:
For the dough:
9 ounces (1 cup plus 2 tablespoons) cold unsalted butter, cut into small pieces
7 1/2 ounces (about 1 2/3 cups) unbleached all-purpose flour
3 3/4 ounces (about 3/4 cup) whole-wheat flour
1/4 cup plus 1/2 tablespoons granulated sugar
1/2 teaspoon kosher salt
1 teaspoon ground cinnamon
For the filling:
6 ounces fresh raspberries (about1 1/2 cups)
3/4 pound small-sized fresh figs (preferably Brown Turkey), quartered (about 2 cups)
3 tablespoons plus 1 teaspoon honey
3 tablespoons plus 1 teaspoon graham cracker crumbs
2 teaspoons finely grated orange zest
1/3 cup plus 2 tablespoons granulated sugar

1 tablespoon fresh thyme, roughly chopped
1 ounce (2 tablespoons) cold unsalted butter, cut into
1 1/2 tablespoons heavy cream

Directions:

For the dough:

Put the flours, cinnamon, sugar, and salt into a food processor. Add the butter; pulse in short bursts until the mixture resembles a coarse meal. Add 3 tablespoons of cold water; pulse again. If the mixture seems dry, add 1 tablespoon of water, pulsing until the mixture forms into a dough. Make sure not to over process. Turn the dough into a clean work surface, gather them together, and then portion into 10 pieces 2 1/2-ounce rounds. Flatten the rounds into disks, individually wrap them in plastic, and refrigerate for at least 2 hours or up to 3 days.

When you are ready to bake, position the racks in the top thirds and the bottom of the oven; preheat the oven to 400F. Line 2 pieces of large-sized, rimmed baking sheets with parchment paper.

Lightly flour a working surface. Place the doughs in the floured surface. With a floured rolling pin, roll each dough into 5 1/2-inch (about 1/8 thick) round disks. Place the round disks into the baking sheets.

For the filling:

In a medium bowl, toss the raspberries, figs, honey, 1/3 cup of the sugar, orange zest, and thyme until combined.

To assemble and bake the crostatas:

Sprinkle each round disk with 1 teaspoon graham cracker crumbs, leaving a 1/2-inch border. The crumbs will soak up the juices that the fruits release during baking, preventing the bottom of the rounds from being soggy. Put a generous 1/4 cup of the filling into the center of each mound, mounding the filling. Top each tart with 1 slice of butter.

Fold the edges of the dough over some of the fruit filling, creating a 1-inch rim and leaving the center exposed. Work your way around, pleating the dough as you go. With a pastry brush, brush the crust of each crostata with cream and then sprinkle the remaining 2 tablespoons sugar over the filling and the crusts.

Bake the crostatas for about 30-35 minutes or until golden brown, rotating and swapping the position of the baking sheets halfway through baking.

Remove the baking sheets from the oven and transfer to racks; let cool for 5 minutes. With an offset spatula, loosen the crostatas and let cool completely on the baking sheets. They are best served on the day they are made.

Nutrition: 450 Calories, 25 g total fat (15 g sat. fat, 6 g mono fat, 1 g poly fat), 5 g protein, 55 g total carbs., 75 g sodium, 65 g Chol., and 4 g fiber.

Pasta Flora or Greek Tart with Apricot Jam

Servings: 6-8
Preparation Time: 15 min
Cooking Time: 45 min
Ingredients:
300 grams apricot jam
3/4 cup sugar
280 grams butter, melted
250 grams whole-wheat flour
250 grams all-purpose flour
2 teaspoons baking powder
2 eggs
Directions:
Whisk the butter, the eggs, and the sugar together. Slowly add the flours and the baking powder, making a soft dough.

Refrigerate the dough for about 30 minutes to rest.

Preheat the oven to 350F or 180C.

Butter well a 25-cm diameter tart pan. Roll out 2/3 of the dough into the buttered tart pan, placing all the way around the raised sides of the pan and gently pressing to evenly cover and join with the base.

Roll out the remaining 1/3 dough into 1/2-cm thickness and then cut into strips.

Spread the jam evenly over the dough in the pan and cover the jam with the strips of dough.

Bake the tart for about 45 minutes.

Nutrition: 656 Calories, 30.2 g total fat (18.4 g sat. fat), 116 mg Chol., 234 mg sodium, 245 mg pot, 91.3 g total carbs, 1.8 g fiber, 35.3 g sugar, and 8.4 g protein.

Frozen Strawberry Greek Yogurt

Servings: 16(1/4 cup)
Preparation Time: 15 min, plus 2-4 hr. freezing
Ingredients:
3 cups Greek yogurt, plain, low-fat (2%)
2 teaspoons vanilla
1/8 teaspoon salt
1/4 cup freshly squeezed lemon juice
1 cup sugar
1 cup strawberries, sliced
Directions:
In a medium-sized bowl, except for the strawberries, combine the rest of the ingredients; whisking until the mixture is smooth. Transfer the yogurt into a 1 1/2 or 2-quart ice cream make and freeze according to the manufacturer's direction, adding the strawberry slices for the last minute. Transfer into an airtight container and freeze for about 2-4 hours. Before serving, let stand for 15 minutes at room temperature.
Nutrition: 86 Calories, 1 g total fat (1 g sat. fat), 3 mg Chol., 16g carbs., 0 g fiber, 15 g sugar, and 4 g protein.

Orange-Sesame Almond Tuiles

Servings: 20 cookies
Preparation Time: 30 min, plus 1 hr. resting
Cooking Time: 45 min
Ingredients:
3/4 cup unblanched or blanched sliced almonds
3 tablespoons orange juice, freshly squeezed
3 tablespoons (about 1 1/2 ounce) unsalted or salted butter
2 tablespoons white sesame seeds
10 tablespoons granulated sugar
1/8 cup whole-wheat flour

1/8 cup all-purpose flour
1 tablespoon toasted sesame oil
1 1/2 teaspoons black sesame seeds
Grated zest of 1 orange, preferably organic
Directions:
In a small-sized saucepan, warm the butter, sesame oil, orange zest, orange juice, and sugar over low heat until the mixture is smooth. Remove from the heat, Stir the flour, almonds and the sesame seeds; let the batter rest for 1 hour at normal room temperature.
Preheat the oven to 375F. Line 2 pieces baking sheet with parchment paper.
Set a rolling pin on a folded dishtowel. Ready a wire rack. Measuring by level tablespoons, drop batter into the prepared baking sheets, placing only 4 on each sheet and spacing them apart evenly.
With dampened fingers, slightly flatten the batter. Place one baking sheet in the oven, bake the tuiles for about 8 to 9 minutes, rotating the baking sheet halfway through baking, until the cookies are evenly browned. Let the cookies cool slightly for 1 minute.
With a metal spatula, lift each cookie of the baking sheet and then drape them over the rolling pin. Let them cool in the rolling pin and then transfer to a wire rack. Repeat the process with the remaining batter. Serve the tuiles a few hours after baking.
Notes: If the cookies cool and harden before you can drape them, they can be softened by putting them back in the oven for about 30-45 seconds. The batter can be made in advance; store in the refrigerator up to 7 days. If not serving immediately, you can store the tuiles in an airtight container to serve later of the same day.
Nutrition: 78 Calories, 4.7 g total fat (1.4 g sat. fat), 5 mg Chol., 13 mg sodium, 39 mg pot., 8.6 g total carbs., 0.7 g fiber, 6.4 g sugar, and 1.1 g protein.

Kataifi
Servings: 8-10
Preparation Time: 50 min

Cooking Time: 30 min
Ingredients:
1-kilogram almonds, blanched and then chopped
1 teaspoon cinnamon
1/4-kilogram kataifi phyllo
2 eggs
4 tablespoons sugar
400 g butter
For the syrup:
1 1/2 kilograms sugar
1 lemon rind
1 teaspoon lemon juice
5 cups water
Directions:
Preheat the oven to 170C.
Put the sugar, eggs, cinnamon, and the almonds in a bowl.
With your fingers, open the kataifi pastry gently. Lay it on a piece of marble and wood. Put 1 tablespoon of the almond mixture in one end and then roll the pastry into a log or a cylinder. Make sure you fold the pastry a little tight, so the filling is enclosed securely. Repeat the process with the remaining pastry and almond mixture.
Melt the butter and put into a baking dish.
Brush the kataifi rolls with the melted butter, covering all the sides.
Place into baking sheets and bake for about 30 minutes.
Meanwhile, prepare the syrup.
Except for the lemon juice, cook the rest of the syrup ingredients for about 5-10 minutes. Add the lemon juice and let cook for a few minutes until the syrup is slightly thick.
After baking the kataifi, pour the syrup over the still warm rolls. Cover the pastry with a clean towel. Let cool as the kataifi absorbs the syrup.
Notes: You can use pistachio nuts if you don't have almonds.
Nutrition: 1085 Calories, 83.3 total fat (24.6 g sat. fat), 119 mg Chol., 248 mg sodium, 759 mg pot., 76.6 g total carbs., 12.7 g fiber, 59.1 g sugar, and 22.6 g protein.

Hazelnut-Orange Olive Oil Cookies

Servings: 6 dozen cookies
Preparation Time: 30 min, plus 1 hr. firming
Cooking Time: 20 min
Ingredients:
5 ounces (1-1/8 cups) whole-wheat flour
5 ounces (1-1/8 cups) unbleached all-purpose flour
3/4 cup plus 2 tablespoons granulated sugar
2 large eggs
2 cups toasted and skinned hazelnuts
1/4 teaspoon table salt
1/2 cup olive oil, extra-virgin
1 teaspoon vanilla extract, pure
1 teaspoon of baking powder
Finely grated zest of 2 medium-sized oranges (about 1 1/2 packed tablespoon)
Directions:
Put the hazelnuts in a food processor; process until finely ground. In a medium bowl, whisk the ground hazelnuts, flours, baking powder, and salt until blended. With a stand or a hand mixer fitted with a paddle attachment, beat the eggs, oil, sugar, orange zest, and vanilla on LOW speed for about 15 seconds or until the sugar is moistened. Increase the speed to HIGH; mix for 15 minutes more or until well combined, the sugar will be dissolved at this point. Add the hazelnut mixture; mix on LOW speed for about 30 to 60 seconds or until the dough has just pulled together.
Divide the dough into 2 portions. Pile one of the doughs on a piece of parchment paper. With the aid of the parchment paper, shape the dough into a 2-inch diameter 11-inch long log. Wrap the parchment around the log, twisting the ends to secure it. Repeat the process with the remaining dough. Refrigerate and chill for about 1 hour or until firm.
Position the oven racks in the lower thirds and the upper position in the oven; preheat the oven to 350F. Line 4 pieces cookie sheets with nonstick baking liners or parchment paper.

Unwrap the logs. Cut the logs into 1/4-inch thick slices. Set them 1-inch apart from each other on the prepared sheets. Place 2 baking sheets in the oven; bake the cookies for about 10 minutes or until the cookies are light golden around the edges and on the bottoms, swapping and rotating the sheets halfway through the baking. Let the cookies cool completely on racks. These can be kept in an airtight container at normal room temperature for up to 7 days.
Notes: you can make the dough logs ahead of time. Freeze them for up to 1 month.
Nutrition: 60 Calories, 4 g total fat (0 g sat. fat, 3 g mono fat, 0 g poly fat), 5 mg Chol., 15 mg sodium, 6 g total carbs., and 0 g fiber.

Greek Cheesecake

Servings: 8-10servings
Preparation Time: 1 hr., 20 min
Cooking Time: 30 min
Ingredients:
4 eggs
250 grams whole-wheat digestive cookies
125 grams butter, melted
1/2 teaspoon cinnamon
1/2 cup sugar
1/2 cup honey
1 teaspoon vanilla extract
1 teaspoon lemon zest
1 kilo white mizithra cheese, fresh or anything similar like ricotta
For the topping:
750 grams black cherries, pitted
2 leaves gelatin
300 grams sugar
Directions:
Process the digestive biscuits in a food processor until crumbled. Add the butter and cinnamon, process again until the mixture is like wet sand in texture. Press the mixture into a 20-cm spring-

form tin, pressing some of the mixture up the sides of the tin to make a ridge. Refrigerate until ready to use.
Preheat the oven to 180C.
With an electric mixer, beat the sugar and the cheese together until creamy. One by one, add in the eggs, the lemon zest, the vanilla extract, and honey. Pour the cheese mixture over the refrigerated biscuit base.
Place the spring-form tin in the oven and with the oven door ajar, bake for 30 minutes or until firm. Remove the cake from the oven and let cool.
Meanwhile prepare the cherries. Place the gelatin leaves in a bowl with cold water; soak until soft. Put the sugar and the pitted cherries into a frying pan, heat over high flame or hear; stew for about 6 minutes or until the cherries release their juices. Add in the softened gelatins; stir well until well mixed. Remove the pan from the heat and let cool for a bit. When slightly cool, pour over the cooled cheesecake.
Refrigerate until the cherry topping set. Serve cold. If desired, serve with vanilla ice cream.
Nutrition: 561 Calories, 19.9 g total fat (11 g sat. fat), 123 mg Chol., 247 mg sodium, 242 mg pot, 80.5 g total carbs, 0.6 g fiber, 54.4 g sugar, and 18.8 g protein.

Phyllo Cups

Servings: 12
Preparation Time: 25 min
Cooking Time: 8 mins
Ingredients:
8 sheets (14x9-inch) frozen phyllo dough, thawed
Nonstick cooking spray
4 teaspoons sugar
For the lemon cheesecake filling:
1 package (8 ounce) cream cheese, softened
3 tablespoons lemon curd
1/3 cup sugar
For the berry-honey filling:
3 ounces cream cheese, softened

1/2 cup whipping cream
1/2 teaspoon vanilla
2 tablespoons honey
Fresh strawberries, sliced (or other berries)
For thee macadamia espresso coconut filling:
1 package (8 ounce) cream cheese, softened
1/3 cup sugar
1/2 cup whipping cream
1 teaspoon espresso powder, instant
1/2 cup toasted coconut
1/4 cup macadamia nuts, finely chopped
Directions:
For the phyllo cups:
Preheat the oven to 350F.
Lightly grease 12 pieces of 2 1/2-inch muffin cups with the cooking spray; set aside.
Lay out 1 phyllo sheet, lightly grease with the cooking spray, sprinkle with some sugar, and then top with another 1 phyllo sheet. Repeat the process until 4 phyllo sheets are stacked, lightly greasing with the cooking spray and sprinkling with the sugar in the process. Repeat the procedure to make 2 stacks of 4-pieces phyllo sheets. Cut each stack lengthwise into halves. Then cut crosswise into thirds; making 12 rectangles.
Press 1 rectangle into each greased muffin cup, pleating the phyllo to form a cup as necessary. Put the muffin cups in the oven and bake for about 8 minutes or until the phyllo cups are golden. When baked, remove the muffin tins from the oven and let cool in the pan for about 5 minutes. Remove the phyllo cups from the muffin tins and let cool completely. Fill each cup with desire filling. They can be filled for up to 1 hour before serving.
For the lemon cheesecake filling:
Put the cream cheese and the sugar into a bowl; beat until the mixture is smooth. Beat in the lemon curd until mixed. Spoon the mixture into phyllo cups. If desired, garnish with lemon peel twists.
For the berry-honey filling:
Put the cream cheese in a bowl; beat until smooth. Beat in the vanilla and the honey. Add in the whipping cream; beat until

stiff peaks form. Spoon the mixture into phyllo cups. Top with sliced strawberries or with preferred berry. Drizzle with more honey, if desired.
For thee macadamia espresso coconut filling:
Put the cream cheese, sugar, and the espresso powder in a bowl; beat. Add in the whipping cream until stiff peaks form Stir in the nuts and toasted coconut. Spoon the mixture into phyllo cups. If desired, garnish with additional toasted coconut and nuts.
Nutrition: 161 Calories, 8 g total fat (4 g sat. fat), 22 mg Chol., 148 mg sodium, 20 g total carbs, 1 g fiber, and 3 g protein.

Poached Cherries

Servings: 5(1/2 cup each)
Preparation Time: 10 min
Cooking Time: 10 min
Ingredients:
1 pound fresh and sweet cherries, rinsed, pitted
3 strips (1x3 inches each) orange zest,
3 strips (1x3 inches each) lemon zest,
2/3 cup sugar
15 peppercorns
1/4 vanilla bean, split but not scraped
1 3/4 cups water
Directions:
In a saucepan, mix the water, citrus zest, sugar, peppercorns, and vanilla bean; bring to a boil, stirring until the sugar is dissolved. Add the cherries; simmer for about 10 minutes until the cherries are soft, but not falling apart. Skim any foam from the surface and let the poached cherries cool. Refrigerate with the poaching liquid. Before serving, strain the cherries.
Nutrition: 170 Calories, 1 g total fat (0 g sat. fat, 0 g mono fat, 0.5 g poly fat), 0 mg Chol., 0 mg sodium, 42 g total carbs., and 2 g fiber.

Watermelon-Strawberry Rosewater Yogurt Panna Cotta

Servings: 4
Preparation Time: 20 min
Cooking Time: 5 min
Ingredients:
500 g seedless watermelon, peeled, and cut into 5-mm pieces
3 teaspoons rosewater
250 ml honey-flavored yogurt
250 ml (1 cup) thickened cream
2 teaspoons gelatin powder
2 tablespoons caster sugar
10 strawberries, washed, hulled, and cut into 5-mm pieces
1 tablespoon hot water
Honey, to serve
Vegetable oil, to grease
Directions:
Brush 4 pieces of 125 ml or 1/2 cup sprinkle molds with vegetable oil to grease.
Put the yogurt into a large-sized heat-safe bowl.
Place the sugar and the cream into a small-sized saucepan and heat over medium heat; stir until the sugar is heated through and the sugar is dissolved.
Place the hot water into a small-sized heat-safe bowl. Sprinkle the gelatin over the hot water. Place the bowl into a small-sized saucepan. Add enough boiling water to fill the saucepan about 3/4 deep on the side of the bowl. With a fork, whisk the mixture until the gelatin is dissolved.
Add the gelatin mixture and the cream mixture into the yogurt, whisking until well combined. Strain the mixture through a fine sieve over a large-sized jug. Pour the strained mixture into the prepared molds. Cover each mold with a plastic wrap.
Refrigerate for at least 6 hours or overnight until set.
In a medium bowl, combine the strawberry, watermelon, and rosewater.

Turn the panna cottas into serving bowl. Spoon the strawberry-watermelon over each panna cotta. Drizzle with honey and serve.

Notes: For a different version, you can omit the rosewater, strawberries, and the honey. Combine the watermelon with 1/3 cup of fresh passion fruit pulp, and spoon over the panna cottas.

Nutrition: 364.96 Calories, 26 g total fat (16 g sat. fat), 75 mg Chol., 69.54 mg sodium, 26 g total carbs., 26 g sugar, 7 g protein, and 1 g fiber.

Mascarpone and Ricotta Stuffed Dates

Servings: 5
Preparation Time: 20 min
Cooking Time: 10 min
Ingredients:
125 g fresh ricotta
125 g mascarpone
2 teaspoons finely grated orange rind
30 pieces fresh dates
45 g (1/4 cup) dry roasted hazelnuts, coarsely chopped, for sprinkling
45 g (1/4 cup) icing sugar mixture
For the Frangelico syrup:
80 ml (1/3 cup) Frangelico liqueur
125 ml (1/2 cup) water
215 g (1 cup) caster sugar
Directions:
With an electric beater, beat the mascarpone, icing sugar, ricotta, and orange rind into a large-sized bowl until the mixture is smooth.

With a sharp knife, cut a slit in each date. Remove the stones and discard. Spoon 1 heaped teaspoon of the ricotta mixture into each date.

To make the Frangelico syrup:
Put the water and the sugar into a medium-sized saucepan. Heat over low heat; cook for about 2 to 3 minutes, stirring until the sugar is dissolved. Increase the heat to high and bring the

mixture to a boil. Cook for 5 minutes without stirring or until the syrup is slightly thick. Stir the Frangelico liqueur. Remove from saucepan from the heat, set aside for 30 minutes to cool. Put the dates into a serving platter. Pour the Frangelico syrup over the dates. Sprinkle with hazelnuts and then serve.
Nutrition: 115.92 Calories, 3.5 g total fat (1.5 g sat. fat), 26 g total carbs., 20 g sugar, 1.5 g protein, and 1 g fiber.

Mediterranean Stuffed Dates

Servings: 24
Preparation Time: 30 min
Ingredients:
24 pieces Medjool dates
1/2 cup (or 8 tablespoons) toasted grated coconut
1/2 cup (or 8 tablespoons) toasted ground pecans
1 mint leaf, for garnish
Directions:
In a small-sized bowl, combine the cheese, grated coconut, and 2 teaspoons of the pecans until well mixed. Transfer the cheese mixture into a plastic bag or a piping bag.
With a sharp knife, cut lengthwise slits across the dates and remove the pits.
If using a plastic bag, cut a small opening in the corner of the plastic bag. Pipe the cheese mixture into the center of the dates. If desired, dip the exposed cheese mixture with the remaining ground pecans.
Arrange the dates into a serving plate; refrigerate and chill for 2 hours. When ready to serve, garnish with mint leaves. This treat is best served cold.
Nutrition: 62 Calories, 4 g total fat (0.8 g sat. fat), 0 mg Chol., 1 mg sodium, 81 mg pot., 7.2 g total carbs., 1.3 g fiber, 5.5 g sugar, 0.8 g protein, 0% vitamin A, 0% vitamin C, 1% calcium, and 3% iron.

Glazed Mediterranean Puffy Fig

Servings: 8

Preparation Time: 5 min
Cooking Time: 25 min
Ingredients:
2 sheets (from 1 pack of 4 sheets) puff pastry
20 figs or dry figs (dry or fresh)
8 ounces mascarpone cheese
2 tablespoons butter
1/2 cup (or 8 tablespoons) honey
1/2 teaspoon cinnamon
1/2 teaspoon nutmeg
1/4 teaspoon salt
4 mint leaves, for garnish
Directions:
Preheat the oven 400F.
Slice the puff pastry into triangle and place into a nonstick baking sheet; bake for about 15-20 minutes or until golden brown. When bakes, remove from the oven and allow to cool.
If using dry figs, rehydrate for 1 hour and then cut into half. Put the butter into a nonstick pan over medium flame or heat. Add the figs; cook for about 3 to 5 minutes. Add the honey, salt, cinnamon, and nutmeg; cook, stirring, for about 3 minutes. Remove the skillet from hat and allow to cool for about 5 to 10 minutes.
Place a baked pastry slice in a serving plate, top with 1 tablespoon of cheese, some figs, and then drizzle with the glaze. Repeat the topping, if desired. Garnish with the mint leaves and serve.
Nutrition: 486 Calories, 22.8 g total fat (8.3 g sat. fat), 22 mg Chol., 226 mg sodium, 391 mg pot., 67.4 g total carbs., 5.4 g fiber, 40.6 g sugar, 7.9 g protein, 5% vitamin A, 1% vitamin C, 14% calcium, and 12% iron.

Mediterranean Stuffed Custard Pancakes

Servings: 10
Preparation Time: 60 min
Cooking Time: 20 min
Ingredients:

For the batter:
- 2 cups flour
- 1/2 cup whole-wheat flour
- 2 cups milk
- 1 cup water
- 1 teaspoon yeast
- 1 teaspoon baking powder
- 1 teaspoon sugar

For the custard:
- 2 cups whole milk
- 2 cups fat-free milk or 2 % milk
- 1 cup heavy cream
- 3 tablespoons sugar
- 1/2 cup cornstarch
- 1/2 cup water
- 7 pieces medium-sized white bread, crust removed
- 1 tablespoon rose water
- 1 tablespoon orange blossom water

For the topping:
- 1 cup pistachio
- 1 tablespoon honey or simple syrup

Directions:

For the custard:

In a medium-sized pot, pour in the milks, heavy cream, cornstarch, and sugar; heat the mixture, stirring. Cut the bread into pieces and add into the pot; stir until the mixture starts to thicken. Add the orange and rose water; stir until the custard is very thick. Remove from the heat and then pour into a bowl; let cool for 1 hour, stirring every 15 minutes. Cover with saran wrap and then refrigerate to completely cool.

For the batter:

Mix all the batter ingredients in a mixing bowl, stirring until well combined; let sit for 20 minutes.

Over medium-low flame or heat, heat a nonstick pan. Pour 1/4 cup-worth of the batter to make a 3-inch diameter pancake; cook for about 30 seconds or until the top of the batter is bubbly and no longer wet and the bottom is golden brown. Transfer into a dish to cool. Repeat the process with the remaining batter.

To assemble:
Take out the bowl of custard from the refrigerator. Transfer the chilled custard into a piping bag.
Fold a pancake together, pinching the edges to make a pocket. Pipe the custard into the pancake pocket, filling it. Repeat the process with the remaining pancakes and custard. Top each filled pocket with the ground pistachio. Refrigerate until ready to serve. To serve, transfer the custard-filled pancakes into a serving plate; drizzle with honey or simple syrup.
Nutrition: 450 Calories, 19 g total fat (8 g sat. fat), 42.9 mg Chol., 241.5 mg sodium, 60 g total carbs., 2.8 g fiber, 16.6 g sugar, 13 g protein, 10.3% vitamin A, 2% vitamin C, 31.2% calcium, and 11.5% iron.

Mediterranean Cheesecake

Servings: 8
Preparation Time: 15 min
Cooking Time: 20 min
Ingredients:
1 package (8 ounces) cream cheese
1/4 cup sour cream
1/2 cup condensed milk, sweetened
5 tablespoons sugar, divided
1 tablespoon vanilla
1 tablespoon orange blossom
1 tablespoon rose water
1 tablespoon orange zest
1 egg
1/2 cup butter
2 cups phyllo dough or kadaifi
1/2 cup toasted coconut
1/2 cup pistachios
1/2 cup simple syrup
Directions:
Preheat the oven to 325F.

With a hand mixer, mix the condensed milk, cream cheese, and the sour cream in a large bowl until well blended. Alternatively, you can blend them until well blended.

Add the orange zest, orange blossom, rose water, vanilla, and sugar, blend for 1 minute. Add in the egg and blend for 30 seconds.

In another bowl, break the kadaifi into pieces. Add 3 tablespoons of the sugar, and the butter, mix until well combined.

Line the bottom and the sides of a cheesecake pan or a muffin tin with the kadaifi mixture.

Pour the cheesecake mixture into the cheesecake pan or muffin tin, filling 80% of the container. Place into the oven and bake for 20 minutes. Remove from the oven and let completely cool before serving.

When completely cool, slice the cake into 8 portions, top with the syrup, pistachio and/or coconut, and glaze with more simple syrup. Serve.

Nutrition: 742 Calories, 43 g total fat (23 g sat. fat), 129.3 mg Chol., 526.5 mg sodium, 78 g total carbs., 2.6 g fiber, 43.1 g sugar, 12 g protein, 24.8% vitamin A, 6.1% vitamin C, 14.3% calcium, and 15.4% iron.

Mediterranean Bread Pudding

Servings: 6
Preparation Time: 10 min, plus 6 hr. chilling
Cooking Time: 20 min
Ingredients:
1/4 of a large-sized lemon, juiced
1/2 cup sugar
8 white bread slices, edges removed, toasted, or more as needed
2 cups Ashta or Lebanese cream, or more as needed
1 1/2 cup simple syrup
1/2 cup shredded coconut, toasted
1/2 cup pine nuts
Directions:

Put the sugar, lemon juice, and water into a thick-bottomed pan. Place the pan on the stove and heat over high flame or heat; bring to a boil, continuously stirring. When boiling, let simmer for 5 minutes, continuously stirring, until the mixture is amber in color, being careful it does not burn and turn bitter.

Choose a metal pan according to your desired size. Immediately pour the caramel into the pan, swirling the pan to spread the caramel evenly.

In a single layer, arrange the toasted bread on top of the caramel layer. Generously pour the simple syrup over the bread and spread with the Ashta. If you are using a small metal pan, repeat the layer of bread, drizzle of caramel, and Ashta. Generously sprinkle with the coconut and the pine nuts. Cover the pan and refrigerate for at least 6 hours or overnight. When chilled, slice into 6 portion sand serve.

Notes: you can decorate this dessert with your preferred choice of topping, such as pistachios, almonds, strawberries, candied orange, etc. You can even layer the ingredients in glasses and ramekins for a fancy presentation.

Nutrition: 619 Calories, 10 g total fat (2.6 g sat. fat), 0 mg Chol., 208.3 mg sodium, 130 g total carbs., 5.3 g fiber, 94.6 g sugar, 5 g protein, 0.2% vitamin A, 53.4% vitamin C, 2.4% calcium, and 7.2% iron.

Chapter 11. Daily snacks

Sushi Appetizer

Preparation Time: 10 minutes
Servings: 4
Ingredients:
1 large cucumber
3 tablespoons cream cheese
½ teaspoon chives
1 teaspoon lime juice
1 oz Feta cheese, crumbled
¼ teaspoon paprika
½ teaspoon ground black pepper
¾ teaspoon minced garlic
Directions:
Trim the ends of cucumber.
With the help of the vegetable slicer make the lengthwise slices from the cucumber.
Make the spread: churn together cream cheese, chopped chives, lime juice, crumbled Feta, paprika, ground black pepper, and minced garlic.
Then spread every cucumber slice with the cream cheese mixture.
Roll the cucumber slices and secure them with the help of the toothpick.
Nutrition: calories 58, fat 4.2, fiber 0.5, carbs 3.7, protein 2.2

Tuna Salad in Lettuce Cups

Preparation Time: 10 minutes
Cooking time: 10 minutes
Servings: 6
Ingredients:
4 Romaine lettuce leaves
8 oz tuna fillet

1 teaspoon balsamic vinegar
½ teaspoon olive oil
1 tablespoon fresh dill, chopped
¼ teaspoon salt
¾ teaspoon chili pepper
1 tomato, chopped
¾ cup Plain yogurt
Directions:
Rub the tuna fillet with salt and chili pepper.
Then drizzle the fish with olive oil.
Bake tuna for 10 minutes at 365F.
Then chill it little and chop.
In the bowl combine chopped tuna, Plain yogurt, tomato, fresh dill, and balsamic vinegar. Mix up well.
Fill the lettuce leaves with the tuna mixture.
Nutrition: calories 152, fat 11.2, fiber 0.3, carbs 3.4, protein 9

Rice Burgers

Preparation Time: 10 minutes
Cooking time: 30 minutes
Servings: 4
Ingredients:
1/3 cup rice
1 cup of water
½ teaspoon salt
2 tablespoons ricotta cheese
1 egg
¼ cup yellow onion, diced
1 teaspoon sunflower oil
½ teaspoon ground black pepper
1 tablespoon wheat flour, whole grain
Directions:
Pour water in a pan. Add rice and salt.
Close the lid and boil rice for 15 minutes or until it will soak all liquid and will be done.
Meanwhile, heat up oil in the skillet.
Add diced onion and roast it until golden brown.

Combine cooked rice with onion.
Add ground black pepper, wheat flour, and egg.
Mix up the mixture. It should smooth but not liquid.
Then make medium size burgers.
Bake the burgers for 10 minutes at 355F.
Top the cooked appetizer with ricotta cheese.
Nutrition: calories 104, fat 3, fiber 0.5, carbs 15.1, protein 3.7

Wheatberry Burgers

Preparation Time: 25 minutes
Cooking time: 15 minutes
Servings:6
Ingredients:
1 cup wheatberry, cooked
2 eggs
¼ cup ground chicken
1 tablespoon wheat flour, whole grain
1 teaspoon Italian seasoning
1 tablespoon olive oil
1 teaspoon salt
Directions:
In the mixing bowl mix up together wheatberry and ground chicken.
Crack eggs in the mixture.
Then add wheat flour, Italian seasoning, and salt.
Mix up the mass with the help of the spoon until homogenous.
Then make burgers and freeze them in the freezer for 20 minutes.
Heat up olive oil in the skillet.
Place frozen burgers in the hot oil and roast them for 4 minutes from each side over the high heat.
Then cook burgers for 10 minutes more over the medium heat.
Flip them onto another side from time to time.
Nutrition: calories 97, fat 5.7, fiber 1.5, carbs 9.2, protein 5.2

Tzatziki

Preparation Time: 10 minutes
Cooking time: 0 minutes
Servings: 4
Ingredients:
1 large cucumber, trimmed
3 oz Greek yogurt
1 teaspoon olive oil
3 tablespoons fresh dill, chopped
1 tablespoon lime juice
¾ teaspoon salt
1 garlic clove, minced
Directions:
Grate the cucumber and squeeze the juice from it.
Then place the squeezed cucumber in the bowl.
Add Greek yogurt, olive oil, dill, lime juice, salt, and minced garlic.
Mix up the mixture until homogenous.
Store tzatziki in the fridge up to 2 days.
Nutrition: calories 44, fat 1.8, fiber 0.7, carbs 5.1, protein 3.2

Kale Wraps with Apple and Chicken

Preparation Time: 10 minutes
Cooking time: 10 minutes
Servings: 4
Ingredients:
4 kale leaves
4 oz chicken fillet
½ apple
1 tablespoon butter
¼ teaspoon chili pepper

¾ teaspoon salt
1 tablespoon lemon juice
¾ teaspoon dried thyme
Directions:
Chop the chicken fillet into the small cubes.
Then mix up together chicken with chili pepper and salt.
Heat up butter in the skillet.
Add chicken cubes. Roast them for 4 minutes.
Meanwhile, chop the apple into small cubes and add it in the chicken.
Mix up well.
Sprinkle the ingredients with lemon juice and dried thyme.
Cook them for 5 minutes over the medium-high heat.
Fill the kale leaves with the hot chicken mixture and wrap.
Nutrition: calories 106, fat 5.1, fiber 1.1, carbs 6.3, protein 9

Tomato Finger Sandwich

Preparation Time: 10 minutes
Cooking time: 0 minutes
Servings: 6
Ingredients:
6 corn tortillas
1 tablespoon cream cheese
1 tablespoon ricotta cheese
½ teaspoon minced garlic
1 tablespoon fresh dill, chopped
2 tomatoes, sliced
Directions:
Cut every tortilla into 2 triangles.
Then mix up together cream cheese, ricotta cheese, minced garlic, and dill.
Spread 6 triangles with cream cheese mixture.
Then place sliced tomato on them and cover with remaining tortilla triangles.
Nutrition: calories 71, fat 1.6, fiber 2.1, carbs 12.8, protein 2.3

Parsley Cheese Balls

Preparation Time: 10 minutes
Cooking time: 1 minute
Servings: 6
Ingredients:
1/3 cup Cheddar cheese, shredded
1 tablespoon dried dill
1 egg, beaten
½ teaspoon salt
2 tablespoons coconut flakes
3 tablespoons sunflower oil
Directions:
 Mix up together shredded cheese with dried dill, salt, and coconut flakes.
 Then add egg and stir carefully until homogenous.
 After this make small balls from the cheese mixture.
 Heat up sunflower oil in the skillet.
 Place cheese balls in the hot oil and roast them for 10 seconds from each side.
 Dry the cooked cheese balls with the help of the paper towel.
Nutrition: calories 105, fat 10.4, fiber 0.2, carbs 0.7, protein 2.6

Layered Dip

Preparation Time: 10 minutes
Cooking time: 0 minutes
Servings: 12
Ingredients:
½ cup hummus
8 tablespoons tzatziki
1 cup tomatoes, chopped
1 cup cucumbers, chopped
1 teaspoon olive oil
1 tablespoon lemon juice

1/3 cup fresh parsley, chopped
1 jalapeno pepper, chopped
Directions:
In the mixing bowl mix up together fresh parsley, lemon juice, olive oil, cucumbers, tomatoes, and chopped jalapeno pepper.
Then make the layer of ½ part of tomato mixture in the casserole mold or glass mold.
Top it with the layer of hummus.
Then add remaining tomato mixture and flatten it well.
Top it with tzatziki and flatten well.
Store the dip in the fridge for up to 3 hours.
Nutrition: calories 49, fat 3.6, fiber 1, carbs 3.3, protein 1.1

Grilled Tempeh Sticks

Preparation Time: 5 minutes
Cooking time: 8 minutes
Servings: 6
Ingredients:
11 oz soy tempeh
1 teaspoon olive oil
½ teaspoon ground black pepper
¼ teaspoon garlic powder
Directions:
Cut soy tempeh into the sticks.
Sprinkle every tempeh stick with ground black pepper, garlic powder, and olive oil.
Preheat the grill to 375F.
Place the tempeh sticks in the grill and cook them for 4 minutes from each side. The time of cooking depends on the tempeh sticks size.
The cooked tempeh sticks will have a light brown color.
Nutrition: calories 88, fat 2.5, fiber 3.6, carbs 10.2, protein 6.5

Sweet Potato Fries

Preparation Time: 10 minutes
Cooking time: 35 minutes
Servings:5
Ingredients:
1 teaspoon Zaatar spices
3 sweet potatoes
1 tablespoon dried dill
1 teaspoon salt
3 teaspoons sunflower oil
½ teaspoon paprika
Directions:
 Pour water in the crockpot. Peel the sweet potatoes and cut them into the fries.
 Line the baking tray with parchment.
 Place the layer of the sweet potato in the tray.
 Sprinkle the vegetables with dried dill, salt, and paprika.
 Then sprinkle sweet potatoes with Zaatar and mix up well with the help of the fingertips.
 Sprinkle the sweet potato fries with sunflower oil.
 Preheat the oven to 375F.
 Bake the sweet potato fries for 35 minutes. Stir the fries every 10 minutes.
Nutrition: calories 28, fat 2.9, fiber 0.2, carbs 0.6, protein 0.2

Italian Style Potato Fries

Preparation Time: 10 minutes
Cooking time: 40 minutes
Servings:4
Ingredients:
1/3 cup baby red potatoes
1 tablespoon Italian seasoning
3 tablespoons canola oil
1 teaspoon turmeric

½ teaspoon of sea salt
½ teaspoon dried rosemary
1 tablespoon dried dill
Directions:
Cut the red potatoes into the wedges and transfer in the big bowl.
After this, sprinkle the vegetables with Italian seasoning, canola oil, turmeric, sea salt, dried rosemary, and dried dill.
Shake the potato wedges carefully.
Line the baking tray with baking paper.
Place the potatoes wedges in the tray. Flatten it well to make one layer.
Preheat the oven to 375F.
Place the tray with potatoes in the oven and bake for 40 minutes. Stir the potatoes with the help of the spatula from time to time.
The potato fries are cooked when they have crunchy edges.
Nutrition: calories 122, fat 11.6, fiber 0.5, carbs 4.5, protein 0.6

Lemon Cauliflower Florets

Preparation Time: 15 minutes
Cooking time: 12 minutes
Servings: 6
Ingredients:
1-pound cauliflower head, trimmed
3 tablespoons lemon juice
3 eggs, beaten
1 teaspoon salt
1 teaspoon ground black pepper
2 cups water, for cooking
3 tablespoons almond butter
1 teaspoon turmeric
Directions:
Place the cauliflower head in the pan.
Add water.

Boil the cauliflower for 8 minutes or until it is tender.
Then cool the vegetable well and separate it onto the florets.
Whisk together beaten eggs, salt, ground black pepper, and turmeric.
Dip every cauliflower floret in the egg mixture.
Toss the almond butter in the skillet and heat it up.
Roast the cauliflower florets for 2 minutes from each side over the medium heat.
When the cauliflower florets are golden brown, they are cooked.
Sprinkle the cooked florets with lemon juice.
Nutrition: calories 103, fat 6.9, fiber 2.9, carbs 6.3, protein 6.1

Greek Style Nachos

Preparation Time: 7 minutes
Cooking time: 0 minutes
Servings: 3
Ingredients:
3 oz tortilla chips
¼ cup Greek yogurt
1 tablespoon fresh parsley, chopped
¼ teaspoon minced garlic
2 kalamata olives, chopped
1 teaspoon paprika
¼ teaspoon ground thyme
Directions:
In the mixing bowl mix up together Greek yogurt, parsley, minced garlic, olives, paprika, and thyme.
Then add tortilla chips and mix up gently.
The snack should be served immediately.
Nutrition: calories 81, fat 1.6, fiber 2.2, carbs 14.1, protein 3.5

Cheesy Phyllo Bites

Preparation Time: 10 minutes
Cooking time: 15 minutes
Servings:8
Ingredients:
3 Phyllo sheets
½ cup Cheddar cheese
2 eggs, beaten
1 tablespoon butter
Directions:
 Mix up together Cheddar cheese with eggs.
 Spread the round springform pan with butter.
 Place 2 Phyllo sheets inside the springform pan.
 Place Cheddar cheese mixture over the Phyllo sheets and cover it with the remaining Phyllo dough sheet.
 Preheat the oven to 365F.
 Cut the Phyllo dough pie onto 8 pieces and bake for 15 minutes.
Nutrition: calories 113, fat 5.4, fiber 0.4, carbs 11.4, protein 5

Cheddar Hot Pepper Dip

Preparation Time: 5 minutes
Cooking time: 10 minutes
Servings:6
Ingredients:
1 cup Cheddar cheese
¼ cup cilantro, chopped
1 chili pepper, chopped
1 teaspoon garlic powder
¼ cup milk
Directions:
 Bring the milk to boil.
 Then add Cheddar cheese in the milk and simmer the mixture for 2 minutes. Stir it constantly.

After this, add cilantro, chili pepper, and garlic powder. Mix up the mixture well. If it doesn't get a smooth texture, use the hand blender to blend the mass.

It is recommended to serve the dip when it gets the room temperature.

Nutrition: calories 83, fat 6.5, fiber 0.1, carbs 1.2, protein 5.1

Traditional Mediterranean Hummus

Preparation Time: 10 minutes
Cooking time: 45 minutes
Servings: 7
Ingredients:
1 cup chickpeas, soaked
6 cups of water
½ cup lemon juice
3 tablespoon olive oil
1 teaspoon salt
1/3 teaspoon harissa
Directions:
Combine chickpeas and water and boil for 45 minutes or until chickpeas are tender.
Then transfer chickpeas in the food processor.
Add 1 cup of chickpeas water and lemon juice.
After this, add salt and harissa.
Blend the hummus until it is smooth and fluffy.
Add olive oil and pulse it for 10 seconds more.
Transfer the cooked hummus in the bowl and store it in the fridge up to 2 days.
Nutrition: calories 160, fat 7.9, fiber 5. carbs 17.8, protein 5.7

Easy Nachos

Preparation Time: 10 minutes

Cooking time: 10 minutes
Servings: 7
Ingredients:
1 cup nachos
1/3 cup Monterey Jack cheese, shredded
2 oz black olives, sliced
2 tomatoes, chopped
Directions:
 Crash the nachos gently and arrange them in the casserole mold in one layer.
 Make the layer of black olives and tomatoes over the nachos. Flatten the ingredients with the help of spatula if needed.
 Then make the layer of cheese and cover casserole mold with foil. Secure the edges.
 Bake the nachos for 10 minutes at 365F.
 Then remove the foil from the mold and serve nachos in the casserole mold.
Nutrition: calories 133, fat 8, fiber 2.3, carbs 10.6, protein 5.4

Salty Almonds

Preparation Time: 1 hour 10 minutes
Cooking time: 15 minutes
Servings: 5
Ingredients:
1 cup almonds
3 tablespoons salt
2 cups of water
Directions:
 Bring water to boil.
 After this, add 2 tablespoons of salt in water and stir it.
 When salt is dissolved, add almonds and let them soak for at least 1 hour.
 Meanwhile, line the tray with baking paper and preheat oven to 350F.

Dry the soaked almonds with a paper towel well and arrange them in one layer in the tray.
Sprinkle buts with remaining salt.
Bake the snack for 15 minutes. Mix it from time to time with the help of the spatula or spoon.
Nutrition: calories 110, fat 9.5, fiber 2.4, carbs 4.1, protein 4

Zucchini Chips

Preparation Time: 15 minutes
Cooking time: 20 minutes
Servings: 4
Ingredients:
1 zucchini
2 oz Parmesan, grated
½ teaspoon paprika
1 teaspoon olive oil
Directions:
Trim zucchini and slice it into the chips with the help of the vegetable slices.
Then mix up together Parmesan and paprika.
Sprinkle the zucchini chips with olive oil.
After this, dip every zucchini slice in the cheese mixture.
Place the zucchini chips in the lined baking tray and bake for 20 minutes at 375F.
Flip the zucchini sliced onto another side after 10 minutes of cooking.
Chill the cooked chips well.
Nutrition: calories 64, fat 4.3, fiber 0.6, carbs 2.3, protein 5.2

Chili Chicken Wings

Preparation Time: 10 minutes
Cooking time: 20 minutes

Servings:3
Ingredients:
3 chicken wings, boneless
1 teaspoon chili pepper, minced
1tablespoon olive oil
1 teaspoon minced garlic
2 tablespoons balsamic vinegar
½ teaspoon salt
Directions:
 Make the chicken sauce: whisk together minced chili pepper, olive oil, minced garlic, balsamic vinegar, and salt.
 Preheat the oven to 360F.
 Line the baking tray with parchment.
 Rub the chicken wings with chicken sauce generously and transfer in the tray.
 Bake the poultry for 20 minutes. Flip them onto another side after 10 minutes of cooking.
Nutrition: calories 138, fat 11, fiber 0.2, carbs 3.8, protein 5.9

Radish Flatbread Bites

Preparation Time: 10 minutes
Cooking time: 10 minutes
Servings:8
Ingredients:
2 tablespoons butter
1/3 cup milk
1 ½ cup wheat flour, whole grain
1 teaspoon salt
1 teaspoon avocado oil
1 cup radish
1 tablespoon cream cheese
Directions:
 Melt butter and combine it together with milk. Stir the liquid.
 Then mix up together flour with butter mixture.
 Knead the soft and non-sticky dough.

Cut the dough into 8 pieces.
Roll up every dough piece into the circle (flatbread).
Pour avocado oil in the skillet.
Roast the flatbreads for 1 minute from each side over the medium heat.
After this, slice the radish and mix it up with cream cheese and salt.
Top cooked flatbreads with radish.
Nutrition: calories 123, fat 3.8, fiber 0.9, carbs 18.9, protein 3

Endive Bites

Preparation Time: 10 minutes
Cooking time: 0 minutes
Servings:10
Ingredients:
6 oz endive
2 pears, chopped
4 oz Blue cheese, crumbled
1 teaspoon olive oil
1 teaspoon lemon juice
¾ teaspoon ground cinnamon
Directions:
Separate endive into the spears (10 spears).
In the bowl combine chopped pears, olive oil, lemon juice, ground cinnamon, and Blue cheese.
Fill the endive spears with cheese mixture.
Nutrition: calories 72, fat 3.8, fiber 1.9, carbs 7.4, protein 2.8

Eggplant Bites

Preparation Time: 15 minutes
Cooking time: 30 minutes
Servings:8

Ingredients:
2 eggs, beaten
3 oz Parmesan, grated
1 tablespoon coconut flakes
½ teaspoon ground paprika
1 teaspoon salt
2 eggplants, trimmed
Directions:
Slice the eggplants into the thin circles. Use the vegetable slicer for this step.
After this, sprinkle the vegetables with salt and mix up. Leave them for 5-10 minutes.
Then drain eggplant juice and sprinkle them with ground paprika.
Mix up together coconut flakes and Parmesan.
Dip every eggplant circle in the egg and then coat in Parmesan mixture.
Line the baking tray with parchment and place eggplants on it.
Bake the vegetables for 30 minutes at 360F. Flip the eggplants into another side after 12 minutes of cooking.
Nutrition: calories 87, fat 3.9, fiber 5, carbs 8.7, protein 6.2

Peanut Butter Yogurt Dip

Preparation Time: 10 minutes
Cooking time: 0 minutes
Servings: 4
Ingredients:
2 tablespoons peanut butter
1 oz Greek Yogurt
1 teaspoon sesame seeds
½ teaspoon vanilla extract
1 tablespoon honey
Directions:
Put peanut butter and Greek yogurt in the big bowl.
With the help of the mixer mix up the mixture until fluffy.

After this, add sesame seeds, vanilla extract, and honey.
Stir it carefully.
Store the dip in the fridge.
Nutrition: calories 74, fat 4.5, fiber 0.6, carbs 6.5, protein 2.9

Roasted Chickpeas

Preparation Time: 10 minutes
Cooking time: 3 hours
Servings: 8
Ingredients:
1 cup chickpeas, canned
1 teaspoon salt
½ teaspoon ground coriander
½ teaspoon ground paprika
½ teaspoon dried thyme
¾ teaspoon cayenne pepper
2 tablespoons olive oil
Directions:
Drain the chickpeas and dry them carefully with the help of the towel.
After this, place them in the baking tray.
Mix up together salt, ground coriander, ground paprika, dried thyme, and cayenne pepper.
Sprinkle the chickpeas with spices and shake well.
After this, drizzle them with olive oil. Give a good shake again.
Preheat the oven to 375F.
Place the tray with chickpeas in the preheated oven and cook them for 35 minutes.
Flip the chickpeas on another side from time to time.
Nutrition: calories 122, fat 5.1, fiber 4.5, carbs 15.4, protein 4.9

Bell Pepper Muffins

Preparation Time: 15 minutes
Cooking time: 15 minutes
Servings: 4
Ingredients:
4 eggs, beaten
4 teaspoons butter, softened
1 teaspoon baking powder
2 bell peppers, chopped
4 tablespoons wheat flour, whole grain
½ teaspoon ground black pepper
½ teaspoon salt
Directions:
 Mix up together eggs, butter, baking powder, wheat flour, ground black pepper, and salt.
 When the batter is smooth, add chopped bell pepper. Stir well.
 Fill ½ part of every muffin mold with bell pepper batter.
 Bake the muffins for 15 minutes at 365F.
Nutrition: calories 146, fat 8.4, fiber 1.1, carbs 11.6, protein 7

Whole-Grain Lavash Chips

Preparation Time: 8 minutes
Cooking time: 10 minutes
Servings: 4
Ingredients:
1 lavash sheet, whole grain
1 tablespoon canola oil
1 teaspoon paprika
½ teaspoon chili pepper
½ teaspoon salt
Directions:

In the shallow bowl whisk together canola oil, paprika, chili pepper, and salt.
Then chop lavash sheet roughly (in the shape of chips).
Sprinkle lavash chips with oil mixture and arrange in the tray to get one thin layer.
Bake the lavash chips for 10 minutes at 365F. Flip them on another side from time to time to avoid burning.
Cool the cooked chips well.
Nutrition: calories 73, fat 4, fiber 0.7, carbs 8.4, protein 1.6

Quinoa Granola

Preparation Time: 10 minutes
Cooking time: 25 minutes
Servings:15
Ingredients:
1 cup rolled oats
6 oz quinoa
7 oz almonds, chopped
5 tablespoons maple syrup
3 tablespoons peanut butter
1 teaspoon ground cinnamon
1 tablespoon coconut flakes
Directions:
In the bog bowl mix up together rolled oats, quinoa, almonds, and coconut flakes.
Then add peanut butter and maple syrup.
Stir the mixture carefully with the help of the spoon.
Line the baking tray with parchment.
Transfer the quinoa mixture in the tray and flatten it well.
Bake granola for 25 minutes at 355F.
Chill the cooked granola well and crack on the servings.
Nutrition: calories 177, fat 9.4, fiber 3.3, carbs 19.1, protein 5.9

Cheesy Artichoke Dip

Preparation Time: 10 minutes
Cooking time: 10 minutes
Servings: 6
Ingredients:
1 cup sour cream
1 cup fresh spinach
4 oz artichoke hearts, drained
1 cup Mozzarella cheese, shredded
1 teaspoon chili flakes
Directions:
Chop the artichoke hearts on the tiny pieces.
Put spinach in a blender and blend until smooth.
Mix up together spinach with artichokes. Add sour cream, Mozzarella cheese, and chili flakes. Stir well.
Transfer the mixture in the mold/pan and flatten it.
Bake the dip for 10 minutes at 360F.
Nutrition: calories 105, fat 8.9, fiber 1.1, carbs 4, protein 3.3

Date and Fig Smoothie

Preparation Time: 5 minutes
Cooking time: 0 minutes
Servings: 1
Ingredients:
1 date, pitted
1 fig, chopped
1 oz Greek yogurt
1/3 cup organic almond milk
1/3 teaspoon ground cardamom
1 teaspoon honey
Directions:
Place all ingredients in the food processor.
Blend the mixture until smooth.
After this, pour the cooked smoothie in the serving glass.

Nutrition: calories 146, fat 3.1, fiber 2.7, carbs 27.7, protein 4.3

Cucumber Bites with Creamy Avocado

Preparation Time: 10 minutes
Cooking time: 0 minutes
Servings: 5
Ingredients:
1 cucumber
5 cherry tomatoes
2 oz avocado, pitted
¼ teaspoon minced garlic
¼ teaspoon dried basil
¾ teaspoon sour cream
¾ teaspoon lemon juice
Directions:
Trim the cucumber and slice it on 5 thick slices.
After this, churn avocado until you get cream mass.
Add minced garlic, dried basil, sour cream, and lemon juice. Mix up well.
Spread the avocado mass over the cucumber slices and top it with cherry tomatoes.
Nutrition: calories 56, fat 2.7, fiber 2.5, carbs 8.1, protein 1.7

Beetroot Chips

Preparation Time: 10 minutes
Cooking time: 15 minutes
Servings: 4
Ingredients:
1 beetroot, peeled
1 teaspoon salt
1 tablespoon sunflower oil
Directions:

Thinly slice the beetroot and sprinkle with salt.
Add the sunflower oil and stir gently with the help of the spatula.
Arrange the beetroot chips in the tray one-by-one and bake for 12 minutes at 370F.
Then flip chips on another side and bake for 3 minutes more.
Nutrition: calories 42, fat 3.6, fiber 0.5, carbs 2.5, protein 0.4

Chapter 12. Eating out

Just because you enjoy eating at restaurants, does not mean you have to ditch the diet. The Mediterranean way of eating positively encourages making meals a social event. It can be a time to get together and unwind. Their way of life might be slower, but there is no reason why you cannot incorporate it into your own new lifestyle. Here are a few tips to help you when eating out:

- As you take a seat, have a glass of water. Studies have shown that drinking 17ounces of water prior to a meal, gives you 44% chance not to overeat, therefore assists in weight loss.
- Avoid breadbaskets. Eat whole-wheat bread at best but save that for home and in moderation.
- Avoid fried foods, unless you are confident, they are cooked in olive oil. The only to find is to ask, if you're bold enough.
- Skip the appetizer or share one at the very least.
- For your main course, chose chicken, or lean pork if you prefer a meat dish. Or consider having fish instead. Better yet, have a vegetarian plate
- Avoid dishes with sauces. Chances are, they have ample sugar and salt to make them palatable. Again, you could ask, but if you are at a chain restaurant, they may not even know the answer as it comes ready made in bulk. That's not a nice reflection!
- Choose plenty of vegetables, even order more as a side dish.
- Avoid salad dressings.
- Fruit for dessert is always better. If you can't resist a pudding; share it with a few friends, this way you only have a couple of spoons.
- Enjoy one glass of red wine, and then drink water for the rest of the meal.
- Chew slowly until all the food is masticated, and easy to swallow.
- Think about the flavors of your food as you chew. Simply said, don't just eat by design- discover the flavors within.
- Sit down and enjoy the food. Appreciate what you taste and consume

- Restaurant portions may be large, so get into the habit of leaving some food on your plate.

Chapter 13. Recipes for special events

Mediterranean Roasted Potatoes

Servings: 6 to 8
Preparation Time: 20 min
Cooking Time: 40 min
Ingredients:
3 to 4 tablespoons Zaatar (homemade version ingredients follow:
3 to 4 tablespoons extra-virgin olive oil
2 to 3 pounds potatoes, cut up, (I used red and purple fingerling potatoes and golden round Buttercream variety)
2 tablespoons Aleppo Pepper, or to taste
1 lemon, juiced
For the homemade Zaatar:
1 tablespoon sea salt
2 tablespoons dried thyme
2 tablespoons sumac
3 tablespoons sesame seeds, toasted
Directions:
Mix all the homemade Zaatar ingredients until well combined. If you want a finer texture, you can ground the ingredients in a coffee grinder. For this recipe, a coarse texture is preferred.
Preheat the oven to 400F.
Cut the potatoes into bite-sized pieces; put into a large bowl and season with the Aleppo Pepper, Zaatar, lemon juice, and the olive oil. Mix until well combined. Transfer the seasoned and oil potatoes into and baking dish and bake for about 40-50 minutes or until the potatoes are golden brown and tender.
Nutrition: 194 Calories, 9.5 g total fat (1.4 g sat. fat), 0 mg Chol., 947 mg sodium, 650 mg pot., 25.5 g total carbs., 4.5 g fiber, 1.9 g sugar, 3.5 g protein, 1% vitamin A, 54% vitamin C, 7% calcium, and 14% iron.

Mediterranean Father's Day Chicken Burgers

Servings: 4

Preparation Time: 20 min
Cooking Time: 10-15 min
Ingredients:
4 pita breads or your favorite burger buns
Greek yogurt or labneh
For the burgers:
1 1/2 pounds ground chicken
5 ounces fresh spinach, wilted, squeezed very dry, and then chopped (I set the leaves in a bowl and microwaved them 20 seconds at a time until wilted)
2 tablespoons fresh dill, minced
2 medium green onions, thinly sliced
2 cloves of garlic, micro planed or very finely minced
1 teaspoon salt
1 tablespoon black pepper
1 egg
1/2 cup panko crumbs, unseasoned
1/2 cup feta cheese crumble
For the tomato relish:
1 cup red onion, finely diced
1/2 cup flat-leaf parsley, minced
1 teaspoon dried oregano
1 teaspoon salt
1/2 teaspoon black pepper
1 tablespoon lemon juice
1 tablespoon extra-virgin olive oil
2 medium tomatoes, seeded, finely diced
Directions:
For the burgers:
Put all the ingredients in a mixing bowl, gently combine until mixed; cover and refrigerate to chill for a couple of hours until the flavors are blended, and the mixture is firm.
For the tomato relish:
Combine all the relish ingredients together; set aside.
To assemble:
Form the firmed chicken mixture into 4 patties.

Heat the grill to medium and grease the grill with oil. Put the patties on the grill and cook until done or the internal temperature reaches 165F.
Spread yogurt on the bottom of each bun or center of each pita. Top the bun or the pita with the burger and then top the burgers with the tomato relish.
Nutrition: 696 Calories, 23.5 g total fat (7.7 g sat. fat), 210 mg Chol., 2001 mg sodium, 1098 mg pot, 54.8 g total carbs., 5.4 g fiber, 6.4 g sugar, 64.9 g protein, 98% vitamin A, 59% vitamin C, 32% calcium, and 44% iron.

Mediterranean Deviled Eggs

Servings: 3
Preparation Time: 15 min
Cooking Time: 8-10 min
Ingredients:
3 eggs, hardboiled
2 sprigs thyme or rosemary or both
12 capers
1 teaspoon goat cheese
1 teaspoon extra virgin olive oil
1 tablespoon pickled onions
1 tablespoon grainy mustard
Black pepper
Directions:
Peel the eggs. In a lengthwise manner, cut into halves, scoop out the egg yolks and transfer into a bowl.
Chop the herbs, leaving a couple of leaves and sprigs intact for garnish later. Add the chopped herbs into the bowl with egg yolks. Add the olive oil, mustard, and goat cheese; mash to combine.
Fill the egg whites with the egg yolk mixture. Top each filled egg half with a sprinkle of pepper, a couple of capers, and a pickled onion or two. Garnish with your preferred herb.
Nutrition: 101 Calories, 7.1 g total fat (2.1 g sat. fat), 165 mg Chol., 1158 mg sodium, 94 mg pot., 3.9 g total carbs., 1.9 g fiber,

0.7 g sugar, 7.1 g protein, 8% vitamin A, 4% vitamin C, 9% calcium, and 20% iron.

Mediterranean Briny Deviled Eggs

Servings: 2 dozen
Preparation Time: 15 min
Cooking Time: 8-10 min
Ingredients:
1 1/2 tablespoons anchovy paste
1 tablespoon Aleppo pepper, plus additional for garnish (or smoked paprika)
1 tablespoon lemon zest
1 tablespoon white wine vinegar
1 teaspoon freshly ground black pepper
1 teaspoon salt
1/3 cup Kalamata olives, pitted, finely chopped
12 eggs, hardboiled
12 olive oil packed anchovies (one 2-ounce tin), halved, for garnish
3/4 cup mayonnaise
Parsley leaves, for garnish
Directions:
Peel the eggs. In a lengthwise manner, carefully slice the eggs into halves. Scoop out the egg yolks and transfer into a mixing bowl. Set the egg whites aside. With a fork, mash the egg yolks until finely crumbled.
Add the mayonnaise, lemon zest, kalamata olives, Aleppo pepper, anchovy paste, and white wine vinegar, mix well using the fork until well combined. Add the salt and pepper; taste and adjust seasoning. If desired, add more Aleppo pepper.
With the fork, vigorously combine the mixture until smooth. With a teaspoon or a piping bag, fill the egg white halves with the egg yolk mixture.
Garnish each with an anchovy half and parsley leaf. Sprinkle with Aleppo pepper; serve.
Nutrition: 70 Calories, 6.4 g total fat (1.2 g sat. fat), 89 mg Chol., 341 mg sodium, 46 mg pot., 2.2 g total carbs., 0.7 g fiber, 0.7 g

sugar, 3.7 g protein, 3% vitamin A, 1% vitamin C, 2% calcium, and 4% iron.

Mediterranean Guacamole

Servings: 4-6
Preparation Time: 10 min
Ingredients:
1 garlic clove, smashed
1/2 cup cherry tomatoes, diced
1/2 cup Kalamata olives, pitted and halved
1/4 cup red pepper, diced
2 avocados
3 pepperoncini, diced
3 tablespoons pepperoncini juice
Crostini or pita chips, for serving
Directions:
With a potato masher, mash the avocadoes in a mixing bowl. Fold in the rest of the ingredients and season with salt and pepper to taste. Serve with crostini or pita chips.
Nutrition: 463 Calories, 42.9 g total fat (8.7 g sat. fat), 0 mg Chol., 308 mg sodium, 1114 mg pot, 22.3 g total carbs., 15.4 g fiber, 2.7 g sugar, 4.7 g protein, 23% vitamin A, 70% vitamin C, 6% calcium, and 14% iron.

Mediterranean tray-baked chicken

Servings: 4
Preparation Time: 15 min
Cooking Time: 40 min
Ingredients:
4 chicken breasts, skin-on
200-gram pack cherry tomatoes
2 teaspoons olive oil, divided
2 red pepper, deseeded and cut into chunks
1/2 of a 150-gram pack garlic and herb soft cheese, full fat
1 red onion, cut into wedges
Handful black olives

Directions:
Heat the oven to 200C, gas to 6, or fan to 180C.
Mix the onion, 1 teaspoon of the olive oil, and the peppers in a large baking tray. Transfer the baking tray on the top shelf of the oven; cook for10 minutes.
Meanwhile, carefully make a pocket between the flesh and the skin of each chicken breast, careful not to completely pull the skin off from the meat. Divide the cheese between each chicken pocket, pushing under the skin. Smooth the skin back down to the flesh and then brush the skin with the remaining 1 teaspoon of olive oil. Season with salt and pepper; transfer into the baking dish. Add the tomatoes and the olives all over and on top of the chicken
Return the tray to the oven; bake for about 25 to30 minutes or until the chicken is cooked and golden. If desired, serve with Mediterranean Roasted Potatoes.
Nutrition: 401 Calories, 21 g total fat (9 g sat. fat), 9 g total carbs., 3 g fiber, 8 g sugar, and 45 g protein.

Pumpkin Flan

Servings: 12
Preparation Time: 30 min, plus 3 hr. chilling
Cooking Time: 20 min
Ingredients:
1 can (15-ounce) pumpkin purée
1 3/4 cups sugar, divided
1 cardamom pod, cracked
1 cup milk
1 teaspoon orange zest
1 teaspoon vanilla extract
2 1/3 cups heavy cream, divided
2 cinnamon sticks
3 whole star anise
5 whole cloves
6 large-sized egg yolks
6 large-sized eggs
1/4 cup water

Directions:
In a large-sized heat-safe bowl, whisk the eggs, the egg yolks, 3/4 cup of the sugar, and the orange zest. Pour the milk, 2 cups of the heavy cream, star anise, cloves, cardamom pod, and cinnamon sticks into a large-sized saucepan; bring to a simmer over medium flame or heat. When simmering, slowly whisk in the egg mixture. Let the mixture steep for 30 minutes. Strain into a heatproof container. Whisk in the pumpkin puree and the vanilla extract. Refrigerate and chill for 3 hours.

In a heavy, small-sized saucepan, stir 1/4 cup water and the remaining 1 cup sugar. Heat over low flame or heat until the sugar is dissolved. Increase the heat and without stirring, bring the syrup to a boil until the deep amber in color, brushing down the sides of the pan with a wet pastry brush occasionally swirling the pan, about 10 minutes. Stir in the remaining 1/3 cup of the heavy cream. The caramel will vigorously bubble.

Divide the caramel mixture between 12 ramekins; refrigerate to chill until set. Divide the custard between the 12 ramekins, pouring over the set caramel. Place the ramekins into a large-sized, oven-safe pan. Carefully pour hot water into the pan, filling until the hot water is halfway up the ramekin sides. Cover the pan with foil.

Bake for about 20 to 25 minutes or until the center of the custard is set. Remove from the oven; remove the foil cover, let cool until warm enough to handle. Refrigerate the flans to chill until cold. When ready to serve, invert the ramekins into plates to dislodge the flans.

Nutrition: 338.3Calories, 20.4 g total fat (11.2 g sat. fat), 60.1 mg sodium, 242.3 mg Chol., 33.8 g total carbs., 0.5 g fiber, 31.1 g sugar, and 6.2 g protein.

Herb-Roasted Turkey

Servings: 8 to 10
Preparation Time: 20 min
Cooking Time: 2 3/4 hr.
Ingredients:

1-piece (12-14 pounds) turkey, neck and giblets removed, at room temperature for 1 hour
1 1/12 tablespoons freshly ground black pepper
1 lemon, quartered
1 onion, medium-sized, quartered
1 orange, quartered
1 tablespoon fresh rosemary, minced
1 tablespoon fresh sage leaves, minced
1 tablespoon fresh thyme leaves, minced
1 tablespoon lemon zest, finely grated
3 tablespoons kosher salt
6 tablespoons (3/4 stick) unsalted butter, at room temperature

Directions:

Preheat the oven to 450F. Place a rack inside a large-sized roasting pan.

With paper towels, pat the turkey dry. Rub the inside and the outside of the turkey with salt and pepper. Put the turkey on the rack in the pan.

In a small-sized bowl, mix the butter, rosemary, lemon zest, thyme, and sage with a fork. Rub the herb mixture on the outside and the inside cavity of the turkey.

Put the lemon quarters, orange quarters, and the onion inside the turkey cavity. Tuck the tips of the wings under the turkey to prevent them from burning during roasting.

Pour 4 cups of water into the roasting pan; place into the oven and roast uncovered for about 30 minutes. After 30 minutes, reduce the oven temperature to 325F. Baste the turkey with the juices in the pan. If needed, add more water into the roasting pan to maintain at least 1/4-inch of liquid in the bottom of the roasting pan. Continue roasting the turkey for a total of 2 3/4 hours, basting every 30 minutes and tenting the turkey with foil if the skin is turning too dark. The turkey is roasted when an instant-read thermometer reads 165F when inserted in the thickest part of the thigh without touching the bone and the juices run clear when the thermometer is removed.

When cooked, transfer the turkey into a serving platter. Tent the bird with foil; let rest for 1 hour before carving.

Nutrition: 640 Calories, 10 g total fat (6 g sat. fat), 1170 mg sodium, 360 mg Chol., 4 g total carbs., 1 g fiber, 2 g sugar, and 123 g protein.

Pistachio Oil Drizzled Robiola, and Pickled Fig Crostini

Servings: 12
Preparation Time: 15 min, plus 30 min softening
Cooking Time: 15 min
Ingredients:
6 dried figs
2 tablespoons sugar
2 tablespoons pistachios, toasted and shelled
12 slices ciabatta bread
1/4 cup extra-virgin olive oil
1/2 cup red wine vinegar
1/4 cup water
Robiola cheese, at room temperature
Directions:
In a saucepan, combine the sugar, red wine vinegar, dried figs, and water; bring the mixture to a simmer. When simmering, remove from the heat; let sit for about 30 minutes or until the figs are soft. When the figs are soft, cut the figs into halves in a lengthwise manner. Alternatively, you can use 6 pieces fresh figs halve d lengthwise.
Crush the pistachios into fine pieces and then combine with the olive oil.
Grill the slices of ciabatta bread.
Spread the cheese over the warm toasted bread slices. Top with each with a fig half and then drizzle with the pistachio oil.
Nutrition: 132.8 Calories, 5.8 g total fat (1 g sat. fat), 120.7 mg sodium, 1.2 mg Chol., 18.44 g total carbs., 1 g fiber, 4.7 g sugar, and 2.8 g protein.

Frascarelli with Mustard Greens and Pecorino

Servings: 4 to 6

Preparation Time: 30 min
Cooking Time: 30 min
Ingredients:
1 bunch (about 6 ounces) mustard greens, stems and center ribs removed; leaves torn into pieces (about 6 cups)
1/4 cup (1/2 stick) unsalted butter
1/4 cup grated Pecorino cheese, or Parmesan cheese
2 cups semolina flour (pasta flour)
Freshly ground black pepper
Kosher salt
Directions:
Into an 8x8x2-inch baking dish, evenly spread the semolina flour.
Pour 1 cup of water into a small-sized bowl and place the bowl beside the baking dish.
Working in quick 4 to 5 batches, gather the thumb and the fingertips of 1 hand together; Dip the fingers into the bowl of water, lift, and splatter the water over the flour in the baking dish. Repeat the process several times until the surface of the flour is sprinkled with ragged water patches about the size of a nickel. Let stand for about 5 seconds until the flour absorbs the water, the flour forming into individual dumplings.
With a slotted spoon or a spatula, turn the dumplings over and coat them with the still dry semolina flour; transfer into a sieve. Over a baking sheet, shake the sieve gently to remove excess flour and then transfer the dumplings into a large-sized rimmed baking sheet.
Remove the process with the remaining water and semolina until all the flour or the water has been used, making more dumplings. There may be leftover semolina or water.
Working in 3 to 4 batches, cook the dumplings for about 30 seconds in a large-sized pot of slowly boiling salted water; gently swirling the water once or twice to prevent the dumplings from sticking. Be sure that the water is not boiling to rapidly. Otherwise, the dumplings may break apart. With a slotted spoon, transfer the cooked dumplings or frascarelli into a large-sized rimmed baking sheet.

Put the butter into a large-sized skillet; heat over medium high flame or heat for about 2 minutes or until foamy and brown bits are forming in the bottom of the pan. Add the frascarelli into the skillet; toss to gently coat with the butter. Add the mustard greens; gently fold just to coat with the butter and slightly wilted; season with salt and pepper to taste. Divide between 4 or 6 bowl and top each serving with grated cheese.
Nutrition: 450 Calories, 14 g total fat (9 g sat. fat), 250 mg sodium, 35 mg Chol., 63 g total carbs., 4 g fiber, 2 g sugar, and 15 g protein.

Mediterranean Lamb Kebabs

Servings: 4
Preparation Time: 30 min
Cooking Time: 40 min
Ingredients:
1-pound ground lamb
2 tablespoons scallions, chopped
2 tablespoons extra-virgin olive oil
2 tablespoons of crème fraiche
12 large-sized shallots; peel, halved lengthwise, and then trim root ends but keep intact
1/3 cup of water
1/2 teaspoon black pepper, freshly ground
1 teaspoon freshly squeezed lemon juice
1 tablespoon parsley, flat leaf, chopped
1 garlic clove, minced
1 1/4 teaspoons salt
1 1/2 teaspoons pomegranate molasses, divided
Warm pita bread, for serving
Directions:
Light an outdoor grill.
In medium-sized bowl, gently mix the ground lamb, garlic, crème fraiche, salt, and the pepper until combined. With moistened hands, roll lamb mixture to form 16 balls.
Into 8 pieces 10-inch or less metal skewers, alternate skewer 3 halves shallots and 2 pieces lamb balls. Brush kebabs with olive

oil. Place on the grill and cook over medium high heat for about 3 minutes, turning once, until the outside of the lamb balls and the shallots are browned but are not cooked all the way through. Transfer the semi-cooked kebabs into very large-sized deep skillet, about 12-14 inches. Add water, the lemon juice, and 1 teaspoon pomegranate molasses to the water; bring the water mixture to a boil. When boiling, cover and gently simmer for about 30 minutes over low flame or heat or until the meatballs are cooked through and the shallots are very tender.
Uncover the skillet; increase heat to high. Add remaining 1/2 teaspoon pomegranate molasses. Continue cooking for 5 minutes more, basting the shallots and the meatballs occasionally until they are glazed.
Transfer kebabs into a serving platter. Drizzle with the remaining sauce from the skillet. Garnish with parsley and scallions. Serve with warmed pita bread.
Nutrition: 379 Calories, 16.9 g total fat (4.8 g sat. fat), 909 mg sodium, 105 mg Chol., 665 mg pot., 21.3 g total carbs., 0.5 g fiber, 1.7 g sugar, 35.1 g protein, 17% vitamin A, 13% vitamin C, 7% calcium, and 23% iron

Marinated Eggplant with Tahini

Servings: 6
Preparation Time: 25 min, plus 2 hr. marinating
Cooking Time: 15-18 min
Ingredients:
3 medium-sized Japanese eggplants
3 cloves garlic, divided
2/3 cup tahini
2 tablespoons oregano, finely chopped
2 tablespoons cilantro, finely chopped
1/2 cup lemon juice, freshly squeezed, divided
1 jalapeño, finely chopped
Extra-virgin olive oil
Freshly cracked black pepper
Kosher salt
Directions:

Preheat the oven to 425F.

Trim off the stalk ends of the eggplants. In a widthwise manner, cut the eggplants into halves. Cut the fatter halves lengthwise into halves, and then cut each half into 3 wedges. Cut the thinner halves lengthwise into halves, and then cut each half into 2 wedges. You will get about 30 pieces of similar-sized slices.

Put the eggplant slices into a large-sized baking sheet. Brush all the sides with olive oil, coating them generously and then generously season with salt and pepper. Roast for about 15-18 minutes or until the eggplant slices are completely soft inside and golden-brown outside.

Meanwhile, prepare the marinade. Whisk the herbs, jalapeno pepper, 4 tablespoons of olive oil, 3 tablespoons of the lemon juice, 1/4teaspoon black pepper, and 1 teaspoon of salt in a medium-sized mixing bowl until the mixture is smooth; add the clove of garlic.

As soon as the eggplants are roasted, immediately transfer them into the marinade; toss gently to completely coat with the marinade and let marinate for at least 2 hours at room temperature.

While the roasted eggplant slices are marinating, prepare the tahini sauce. Whisk together 2/3 cup of water, the remaining 5 tablespoons of lemon juice, the tahini, 1/2 teaspoon salt, and 2 pieces minced garlic until the mixture is smooth.

To serve, arrange the marinated roasted eggplant slices in a serving platter and then drizzle with the tahini sauce.

Notes: If you are not going to serve the dish on the same day, you can refrigerate the marinated eggplant slices for up to 2 days. When ready to serve, get them out of the fridge and let them come to room temperature for about 1 hour. Make the tahini sauce and serve according to directions.

Nutrition: 400 Calories, 33.8 g total fat (4.9 g sat. fat), 0 mg Chol., 68 mg sodium, 801mg pot., 23.8 g total carbs., 13 g fiber, 8.9 g sugar, 7.7 g protein, 3% vitamin A, 30% vitamin C, 17% calcium, and 21% iron

Chicken and Mediterranean Tabbouleh

Preparation Time: 30 min
Chill: 4 h to 24 hrs.
Stand: 30 min
Ingredients:
6 ounces chicken breast halves, skinless, boneless, broiled or grilled, then sliced
4 large leaves romaine and/or butter head (Bibb or Boston) lettuce
3/4 cup water
3 tablespoons lemon juice
2 tablespoons olive oil
2 tablespoons green onions, thinly sliced
1/8 teaspoon ground black pepper
1/4 teaspoon salt
1/4 cup bulgur
1/2 cup tomatoes (1 medium), chopped
1/2 cup seeded cucumber, finely chopped
1/2 cup Italian parsley, finely chopped
1 tablespoon fresh mint, snipped (or 1 teaspoon dried mint, crushed)
Directions:
In a large sized bowl, combine the bulgur and the water; let stand for 30 minutes. After 30 minutes, drain in the sink through a fine sieve; pressing out the excess water from the bulgur using a large spoon. Return the bulgur into the bowl. Stir in the cucumber, tomatoes, green onions, parsley, and the mint. Prepare the dressing; put the olive oil, lemon juice, salt, and pepper into a screw-top jar. Cover securely and shake well until well mixed. Pour the dressing over the bulgur mixture; lightly toss to coat the bulgur mixture with the dressing. Cover the bowl and refrigerate to chill for at least 4 hours up to 24 hours, occasionally stirring.
When ready to serve, bring the bulgur mixture to room temperature. Divide the romaine and/or butterhead lettuce leaves between 2 shallow bowls, top with the broiled or grilled chicken, and then top with the bulgur mixture.

Notes: To grill prepare the breast halves, lightly sprinkle the chicken breasts with salt and pepper. If using charcoal grill, place on the rack of an uncovered grill and cook over medium heat for about 12-15 minutes, turning once, until the meat is no longer pink and the internal temperature is175F. If using a gas grill, preheat the grill. When the grill is heated, reduce the heat to medium. Place the chicken on the rack, cover, and grill as directed.

Nutrition: 294 Calories, 13 g total fat (2 g sat. fat,2 g poly. fat, 8 g mono. fat), 72 mg Chol., 276 mg sodium, 16 g total carbs., 5 g fiber, 3 g sugar, and 30 g protein.

St. Valentine's Mediterranean Pancakes

Servings: 2
Preparation Time: 2 min
Cooking Time: 20 min
Ingredients:
4 eggs, preferably organic
2 pieces banana, peeled and then cut into small pieces
1/2 teaspoon extra-virgin olive oil (for the pancake pan)
1 tablespoon milled flax seeds, preferably organic
1 tablespoon bee pollen, milled, preferably organic
Directions:
Crack the eggs into a mixing bowl. Add in the banana, flax seeds, and bee pollen. With a hand mixer, blend the ingredients until smooth batter inn texture.

Put a few drops of the olive oil in a nonstick pancake pan over medium flame or heat. Pour some batter into the pan; cook for about 2 minutes, undisturbed until the bottom of the pancake is golden and can be lifted easily from the pan. With a silicon spatula, lift and flip the pancake; cook for about 30seconds more and transfer into a plate.

Repeat the process with the remaining batter, oiling the pan with every new batter.

Serve the pancake as you cook or serve them all together topped with vanilla, strawberry, pine nuts jam.

Nutrition: 272 Calories, 11.6 g total fat (3 g sat. fat), 327 mg Chol., 125 mg sodium, 633 mg pot., 32.7 g total carbs., 4.5 g fiber, 17.3 g sugar, 13.3 g protein, 10% vitamin A, 20% vitamin C, 6% calcium, and 12% iron

Chapter 14. Bonus: Recipes for Air Fryer

Air Fried Dragon Shrimp

Cooking Time: 15 minutes
Servings: 2

Ingredients:

½ lb. shrimp
¼ cup almond flour
Pinch of ginger
1 cup chopped green onions
2 tablespoons olive oil
2 eggs, beaten
½ cup soy sauce

Directions:
Boil the shrimps for 5-minutes. Prepare a paste made of ginger and onion. Now, beat the eggs, add the ginger paste, soya sauce and almond flour and combine well. Add the shrimps to the mixture then place them in a baking dish and spray with oil. Cook shrimps at 390°Fahrenheit for 10-minutes.
Nutrition: Calories: 278, Total Fat: 8.6g, Carbs: 6.2g, Protein: 28.6g

Lasagna Zucchini Cups

Cooking Time: 25 minutes
Servings: 6

Ingredients:
Chopped parsley, for garnish
¼ cup parmesan, freshly grated

½ cup mozzarella, shredded
½ cup ricotta
1-14.5-ounce can have crushed tomatoes
Black pepper and salt to taste
½ teaspoon oregano, dried
½ lb. ground beef
2 garlic cloves, minced
½ onion, chopped
1 tablespoon olive oil
3 zucchinis

Directions:
In a large pan over medium heat, add the oil. Add onion and garlic and cook for 5-minutes. Add in the ground beef and cook for 10-minutes stirring often. Season with oregano, salt, pepper, cook until meat is no longer pink. Add crushed tomatoes and simmer mixture for 5-minutes. Stir in the ricotta and remove from heat. Cut zucchini in half crosswise in two. Using a spoon scoop out zucchini flesh to create wells. Fill wells with meat mixture. Top with mozzarella and parmesan cheese. Place directly in air fryer and cook at 350°Fahrenheit for 15-minutes. Garnish with parsley and parmesan.

Spinach Artichoke Stuffed Peppers

Cooking Time: 15 minutes
Servings: 4

Ingredients:
4 assorted bell peppers, halved and seeded
Salt and black pepper to taste
Olive oil for drizzling
2 cups shredded rotisserie chicken
Fresh parsley, chopped for garnish
2 cloves garlic, minced

¼ cup mayonnaise
¼ cup sour cream
½ cup mozzarella, shredded, divided
6-ounces cream cheese, softened
1 (10-ounce) package frozen spinach, thawed, well-drained, and chopped
1 (14-ounce) can artichoke hearts, drained and chopped

Directions:
On a large, rimmed baking sheet, place bell peppers cut side-up and drizzle with olive oil, then season with salt and pepper. In a large bowl, combine chicken, artichoke hearts, spinach, cream cheese, ½ cup mozzarella, parmesan, sour cream, mayo and garlic. Season with more salt and pepper and mix until well blended. Divide the chicken mixture between pepper halves, top with remaining mozzarella, and bake in air fryer at 400°Fahrenheit for 15-minutes. Garnish with parsley and serve.
Nutrition: Calories: 284, Total Fat: 13.4g, Carbs: 9.2g, Protein: 34.3g

Cauliflower Shepherd's Pie

Cooking Time: 43 minutes
Servings: 4

Ingredients:

1 medium head of cauliflower, cut into florets
¼ cup whole milk
3-ounces cream cheese, softened
1 tablespoon parsley, chopped for garnish
2/3 cup chicken broth
2 tablespoons almond flour
1 cup frozen peas
1 lb. ground beef

2 cloves garlic, minced
2 carrots, peeled, and chopped
1 large onion, chopped
1 tablespoon olive oil
Salt and black pepper to taste

Directions:
Make mashed cauliflower. Bring a pot of water to boil, add the florets and cook for 10-minutes. Drain pot and then use paper towel to absorb excess water. Return florets to pot and mash with potato masher until smooth. Stir in cream cheese, milk and season with salt and pepper. Set aside. Make the beef mixture: in large pan over medium heat, heat oil. Add onion, garlic and cook for 5-minutes.
Add ground beef for 5-minutes or until meat is no longer pink. Stir in frozen peas and corn and cook another 3-minutes. Sprinkle meat mixture with almond flour and stir to even distribute. Cook for another minute then add chicken broth. Bring to a simmer and let mixture thicken slightly, for 5-minutes. Place beef mixture in air fryer baking dish. Top beef mixture with an even layer of cauliflower and bake in air fryer at 400°Fahrenheit for 15-minutes. Garnish with parsley and serve.
Nutrition: Calories: 279, Total Fat: 13.2g, Carbs: 10.2g, Protein: 34.2g

Coconut Lime Skirt Steak

Cooking Time: 5 minutes
Servings: 2

Ingredients:

½ cup coconut oil, melted
Zest of one lime
2-1lb. grass fed skirt steaks

¾ teaspoon sea salt
1 teaspoon red pepper flakes
1 teaspoon ginger, fresh, grated
1 tablespoon garlic, minced
2 tablespoons freshly squeezed lime juice

Directions:
In a mixing bowl, combine lime juice, coconut oil, garlic, ginger, red pepper, salt, and zest. Add the steaks and toss and rub with marinade. Allow the meat to marinate for about 20-minutes at room temperature. Transfer steaks to your air fryer directly on the rack. Cook steaks in air fryer at 400°Fahrenhet for 5-minutes.
Nutrition: Calories: 312, Total Fat: 12.3g, Carbs: 6.4g, Protein: 42.1g

Spicy Chicken Enchilada Casserole

Cooking Time: 40 minutes
Servings: 2

Ingredients:

1 lb. of chicken breasts, skinless and boneless
Salt and pepper to taste
½ cup cilantro, fresh, minced
Olive oil spray
2 cups cheddar cheese, shredded
Lime wedges (optional)
Sour cream (optional)
1 (4-ounce) can of green chilies, chopped
1 cup feta cheese, finely crumbled
1 ½ cup enchilada sauce

Directions:
Pat the chicken breasts dry and season with salt and pepper. Combine the chicken and enchilada sauce in a pan and simmer for 15-minutes over medium-low heat. Flip chicken over and cover and cook for an additional 15-minutes. Remove the chicken from pan and shred into bite-size pieces. Combine shredded chicken, feta cheese, enchilada sauce, chiles, and cilantro in a bowl. Add salt and pepper. Spray the air fryer baking dish with olive oil. Coat the entire bottom and sides. Evenly spread a cup of shredded cheese on the bottom of baking dish. Add the chicken mixture, then add another cup of cheese on top. Bake in your air fryer at 350°Fahrenheit for 10-minutes. Serve with optional lime wedges and sour cream.
Nutrition: Calories: 338, Total Fat: 12.3g, Carbs: 8.3g, Protein: 32.2g

Kale & Ground Beef Casserole

Cooking Time: 16 minutes
Servings: 4
Ingredients:

4-ounces mozzarella, shredded
2 cups marinara sauce
10-ounces kale, fresh
1 teaspoon oregano
1 teaspoon onion powder
½ teaspoon sea salt
1 lb. lean ground beef
2 tablespoons olive oil

Directions:
In a deep skillet, heat the olive oil for 2-minutes, add in the ground beef and cook for an additional 8-minutes or until meat

is browned. Stir in salt, pepper, garlic powder, onion powder and oregano. In batches, stir the kale into beef mixture, cooking for another 2-minutes. Stir in the marinara sauce and cook for 2-minutes more. Mix in half the cheese into mixture. Transfer mixture into the air fryer baking dish. Sprinkle the remaining cheese on top. Broil in air fryer at 400°Fahrenheit for 2-minutes. Allow to rest for 5-minutes before serving.
Nutrition: Calories: 312, Total Fat: 13.2, Carbs: 9.2g, Protein: 43.2g

Cauliflower-Cottage Pie

Cooking Time: 40 minutes
Servings: 4
Ingredients:

Half a cup of bacon bits
2 cups cauliflower rice
¼ cup tomato puree
1 tablespoon coconut oil
½ white onion, chopped
2lbs. lean ground beef
1 tablespoon mixed spice blend

Directions:
In the frying pan add coconut oil and onions cook for 2-minutes. Add the ground beef into pan and cook for an additional 5-minutes or until meat is browned. Add spices and stir to combine. Add the tomato puree and mix well and cook for another 10-minutes. Transfer to air fryer baking dish. Top with cauliflower rice and bacon bits. Bake in air fryer at 350°Fahrenheit for 20-minutes. Serve warm.
Nutrition: Calories: 367, Total Fat: 13.4g, Carbs: 11.2g, Protein: 43.1g

Roasted Asian Shrimp & Brussels Sprouts

Cooking Time: 19 minutes
Servings: 4
Ingredients:

1 lb. jumbo frozen shrimp, thawed and drained
1 lb. brussels sprouts
2 tablespoons olive oil
Salt and pepper to taste

Asian Marinade Sauce:

2 tablespoons rice vinegar
2 tablespoons Splenda
2 teaspoons liquid stevia
1 tablespoon Asian sesame oil
½ teaspoon garlic powder
1/3 cup soy sauce

Directions:
About 20-minutes before you start cooking, place the shrimp into a colander and place it in the sink and let shrimp drain. Mix the soy sauce, stevia, rice vinegar, sesame oil, and garlic powder to make marinade mixture. After the shrimp have drained well, layer them on a paper towel a blot dry, so they are as dry as you can get them. Place dried shrimp into Ziploc bag with half of the marinade and allow the shrimp to marinate while you cook the brussels sprouts. Trim the stem ends off each brussels sprout and cut in half. Place brussels sprouts into a bowl and toss with desired amount of olive oil, salt and pepper.
Spread brussels sprouts out in a single layer in your air fryer and roast at 400°Fahrenheit for 15-minutes. Keep brussels sprouts

in air fryer and move to one side. Add shrimp beside them in air fryer. Roast for an additional 4-minutes. Remove from air fryer and place into serving bowl, add remaining marinade into bowl and give dish a stir. Serve immediately.

Nutrition: Calories: 267, Total Fat: 11.2g, Carbs: 8.3g, Protein: 9.2g

Grilled Chicken with Garlic Sauce

Cooking Time: 15-minutes
Servings: 4
Ingredients:

1lb. chicken breast, cut into large cubes
2 bell peppers, chopped
1 zucchini
1 onion, chopped

For Garlic Sauce:

1 head garlic, peeled
¼ cup lemon juice
1 cup olive oil
1 teaspoon salt

Additional ingredients for the marinade:

1 teaspoon salt
½ cup olive oil

Directions:

Soak 4 wooden skewers in water. For your garlic sauce, place garlic cloves and salt into blender. Then, add in about 1/8 of a cup of lemon juice and ½ a cup of olive oil. Blend for about 10-seconds. Keep half of the garlic sauce to serve with. Take the other half of garlic sauce and add an additional ½ cup of olive oil and a teaspoon of salt and mix well—this will make your marinade. Chop up the chicken, onion, bell peppers, and zucchini into 1-inch cubes or squares. Mix them in a bowl with the marinade. Place the cubes onto the skewers and cook them directly on the air fryer rack at 400°Fahrenheit for 15-minutes. Serve warm.
Nutrition: Calories: 321, Total Fat: 12.5g, Carbs: 9.2g, Protein: 32.1g

Bacon-Wrapped Stuffed Zucchini Boats

Cooking Time: 15 minutes
Servings: 4
Ingredients:

½ a teaspoon of fresh ground black pepper
1 teaspoon sea salt
5-ounces cream cheese
8-mushrooms, finely chopped
1 tablespoon Italian parsley, chopped
1 tablespoon finely chopped dill
3 garlic cloves, peeled, pressed
1 sweet red pepper, finely chopped
2 large zucchinis
12 bacon strips
1 medium onion, chopped

Directions:
Preheat your air fryer to 350°Fahrenheit. Trim the ends off zucchini. Cut zucchini in half lengthwise. Scoop out pulp, leaving ¼-inch thick shells. Stir pulp in mixing bowl. Add

onion, garlic, herbs, pepper, cream cheese, salt, and pepper. Mix well to combine. Fill individual shells with the same amount of stuffing. Wrap three bacon strips around each zucchini boat such that the ends end up underneath. Place them directly on the air fryer rack and bake turning the temperature up to 375°Fahrenheit for 15-minutes. Remove and serve immediately.

Nutrition: Calories: 282, Total Fat: 9.1g, Carbs: 6.3g, Protein: 24.2g

Parmesan Chicken Wings

Cooking Time: 22 minutes
Servings: 4
Ingredients:

2 lbs. chicken wings
2 tablespoons olive oil
1 teaspoon sea salt
1 teaspoon black pepper
3 tablespoons butter
3 tablespoons olive oil
3 garlic cloves, minced
4 tablespoons parmesan cheese
1/8 teaspoon smoked paprika
¼ teaspoon red pepper flakes
Salt and pepper to taste

Directions:
Add chicken to a bowl and pat the chicken dry. Drizzle with 2 tablespoons of olive oil, 1 teaspoon of sea salt, and 1 teaspoon black pepper. Gently toss to coat chicken. Place chicken wings into air fryer directly on the rack. Bake at 400°Fahrenheit for 20-minutes, flipping wings half-way through cook time. In a pan over medium heat add butter and 3 tablespoons olive oil and melt the butter down, for about 3-minutes. Add 2

tablespoons of parmesan cheese, smoked paprika, red pepper flakes, salt and pepper to taste. Cook sauce for about 2-minutes. Remove the wings from air fryer and place in large bowl. Pour the garlic parmesan sauce over the wings toss to coat. Serve wings topped with additional a2 tablespoons of parmesan cheese.

Nutrition: Calories: 324, Total Fat: 12.3g, Carbs: 9.3g, Protein: 39.3g

Beef Burgers

Cooking Time: 10 minutes
Servings: 4

Ingredients:

1 lb. ground beef
1 teaspoon parsley, dried
½ teaspoon oregano, dried
½ teaspoon ground black pepper
½ teaspoon salt
½ teaspoon onion powder
½ teaspoon garlic powder
1 tablespoon Worcestershire sauce
Olive oil cooking spray

Directions:
In a mixing bowl, mix the seasonings. Add the seasoning to beef in a bowl. Mix well to combine. Divide the beef into four patties, put an indent in the middle of patties with your thumb to prevent patties from bunching up in the middle. Place burgers into air fryer and spray the tops of them with olive oil. Cook for 10-minutes at 400°Fahrenheit, no need to flip patties. Serve on a bun with a side dish of your choice.

Nutrition: Calories: 312, Total Fat: 11.3g, Carbs: 7.2g, Protein: 39.2g

Bacon Wrapped Avocado

Cooking Time: 10 minutes
Servings: 2
Ingredients:
2 avocados, fresh and firm
Chili powder
Ground cumin
4 thick slices of hickory smoked bacon

Directions:
Slice the avocados into wedges and peel off the skin. Stretch the bacon strips this will help to elongate them. Slice avocados in half. Next, take half a bacon strip and wrap one around each avocado wedge and tuck the ends under the bottom. Sprinkle wedges with chili powder and cumin. Bake the bacon wrapped avocado wedges in air fryer at 400°Fahrenheit for 10-minutes. Serve with your favorite salad!
Nutrition: Calories: 276, Total Fat: 7.3g, Carbs: 6.3g, Protein: 21g

Buffalo Chicken Meatballs

Cooking Time: 20 minutes
Servings: 4

Ingredients:

1 lb. ground chicken
1 egg, beaten
1 celery stalk, trimmed and finely diced
1 cup buffalo wing sauce

1 teaspoon black pepper
1 teaspoon pink sea salt
1 teaspoon garlic powder
1 teaspoon onion powder
1 tablespoon mayonnaise
1 tablespoon almond flour
2 sprigs of green onion, finely chopped

Directions:
Place the baking pan in air fryer and spray with olive oil. In a bowl, combine all ingredients, except buffalo sauce. Mix well. Use your hands to form 2-inch balls. Place the meatballs in air fryer and bake at 350°Fahrenheit for 15-minutes. Remove the meatballs from the air fryer. Add them to a pan over medium-low heat. Coat meatballs with buffalo sauce and stir cooking in pan for 5-minutes. Serve.
Nutrition: Calories: 302, Total Fat: 12.4g, Carbs: 7.6g, Protein: 32.1g

Caprese Grilled Chicken with Balsamic Vinegar

Cooking Time: 20 minutes
Servings: 6

Ingredients:

6 grilled chicken breasts, boneless, skinless
6 large basil leaves
6 slices of tomato
6 slices of mozzarella cheese
1 tablespoon butter
¼ cup balsamic vinegar

Directions:

Prepare chicken in air fryer at 400°Fahrenheit for 15-minutes or until chicken is cooked. As chicken is cooking, pour balsamic vinegar into the pan and cook until reduced by half, for about 5-minutes. Add in the butter and stir with a flat whisk until well combined. Set aside. Top chicken with mozzarella cheese slices, basil leaves, and tomato slice each. Drizzle with balsamic reduction and serve warm.
Nutrition: Calories: 289, Total Fat: 11.3g, Carbs: 7.2g, Protein: 28g

Philly Cheese Steak Stuffed Peppers

Cooking Time: 40 minutes
Servings: 2

Ingredients:

8-ounces of roast beef, thinly sliced
8-slices of provolone cheese
2 large green bell peppers
1 medium sweet onion, diced
1 (6-ounce) package of baby Bella mushrooms
1 tablespoon garlic, minced
2 tablespoons olive oil
2 tablespoons butter

Directions:
Cut your peppers in half lengthwise, removing ribs and seeds. Slice onions and mushrooms. Sauté over medium heat with butter, olive oil, a dash of salt, pepper, and minced garlic. Cook for 20-minutes or until the mushrooms and onions are sweet and caramelized. Slice the roast beef into thin strips and add to the onion/mushroom mixture. Allow cooking for 10-minutes. In the inside of each pepper line it with a slice of provolone cheese. Fill each pepper with meat mixture. Garnish

top of each pepper with another slice of provolone cheese. Bake in the air fryer at 375°Fahrenhiet for 10-minutes.
Nutrition: Calories: 298, Total Fat: 11.5g, Carbs: 8.2g, Protein: 39.2g

Parmesan, Garlic, Lemon Roasted Zucchini

Cooking Time: 10 minutes
Servings: 4

Ingredients:

1 ½ lbs. zucchini (about 4 small zucchini)
Salt and pepper to taste
¾ cup parmesan cheese, finely shredded
2 cloves garlic, minced
Zest of 1 lemon
2 tablespoons olive oil

Directions:
Cut zucchini into thick wedges or halves (cut each zucchini in half then that half in half, so you have 4 wedges from each zucchini. In a bowl, stir olive oil, garlic, and lemon zest. Align zucchini in air fryer space them evenly apart. Brush olive oil mixture over tops of zucchini. Sprinkle tops with parmesan cheese and season lightly with salt and pepper. Bake in air fryer at 375°Fahrenheit for 10-minutes. Serve warm.

Chicken Filet Stuffed with Sausage

Cooking Time: 15 minutes
Servings: 4

Ingredients:

4 chicken fillets
4 sausages, casings removed

Directions:
Place the sausage inside the chicken filets and roll the fillets. Seal with 2 toothpicks each. Air fry the chicken filets at 375°Fahrenheit for 15-minutes. Serve warm.
Nutrition: Calories: 276, Total Fat: 12.2g, Carbs: 8.2g, Protein: 28g

Bourbon Chicken

Cooking Time: 22 minutes
Servings: 4

Ingredients:

3lbs. of chicken wings
¾ cups ketchup
¼ teaspoon cayenne
¼ cup Bourbon
2 teaspoons smoked paprika
½ cup water
2 garlic cloves, crushed
¼ cup onion, minced
2 teaspoons stevia
1 tablespoon liquid smoke
1 teaspoon salt
½ teaspoon black pepper

Directions:

In a bowl, mix liquid smoke, onion, garlic, ketchup, stevia and cook for 5-minutes in an electric pressure cooker on Sauté function. Combine the rest of your ingredients and cook on high pressure for 5-minutes. Do a quick release of pressure and then transfer the chicken wings into the air fryer basket. Cook wings in air fryer for 6-minutes at 400°Fahrenheit. Dip wings into sauce and air fry for another 6-minutes. Serve hot!
Nutrition: Calories: 302, Total Fat: 12.5g, Carbs: 8.4g, Protein: 32.4g

Roasted Chicken Legs

Cooking Time: 35 minutes
Servings: 2
Ingredients:

2 chicken legs
2 teaspoons sweet smoked paprika
1 teaspoon honey
Salt and pepper to taste
½ teaspoon garlic powder
Fresh parsley, chopped for garnish
1 lime sliced for garnish

Directions:
Combine all the ingredients except the chicken in a bowl. Rub the mixture over the chicken and preheat your air fryer for 3-minutes. Cook the chicken in air fryer at 390°Fahrenheit for 35-minutes. Serve with a favorite salad of your choice.
Nutrition: Calories: 232, Total Fat: 9.3g, Carbs: 7.5g, Protein: 22.1g

Mongolian Chicken

Cooking Time: 17 minutes
Servings: 4

Ingredients:

4 chicken breasts, boneless, skinless, chopped small pieces
1 yellow onion, thinly sliced
Olive oil for frying
1 Chili Paid, chopped
3 garlic cloves, minced
5 curry leaves
1 teaspoon ginger, grated
¾ cup evaporated milk

Marinade:

1 egg
1 tablespoon light soy sauce
Self-rising flour to coat
½ tablespoon cornstarch
Seasonings:

1 teaspoon liquid stevia
1 tablespoon chili sauce
½ teaspoon sea salt
Dash of black pepper

Directions:
Combine all of you marinade ingredients in a bowl and marinate the chicken with it for an hour. Dredge the chicken in the self-rising flour and spray some oil over. Cook in air fryer for 10-minutes at 390°Fahrenheit. Heat a wok and sauté the ginger, garlic, chili paid, curry leaves and onions for 2-minutes. Add the chicken and seasonings, stirring to combine well. Add your milk and cook until thickened. Serve hot!

Nutrition: Calories: 286, Total Fat: 11.3g, Carbs: 6.4g, Protein: 28g

Chapter 15. Bonus: Traditional Italian recipes

Preparation Time: 18 Minutes
Servings: 4

Ingredients Needed
Tomato paste – 2 tablespoons
Cherry tomatoes – 12 (quartered)
Black olives – 1/3 cup (sliced)
Lavash bread – 4 sheets
Thinly sliced salami – 1 (4 ounce) package
Small red onion – 1 (diced)
Mozzarella cheese – 1 ¾ cups (shredded)
Instructions to follow
Set the oven rack about 6 inches from the heat source and preheat the oven's broiler and line two baking sheets with aluminum foil.
Place on each baking sheet, two sheets of lavash and spread over each sheet, 1 ½ teaspoon of tomato paste. Lay atop the tomato paste, 6 pieces of salami each. After this, scatter cherry tomatoes, red onions, and black olives on top of each, then cover with mozzarella cheese.
Broil lavash in batches in the preheated oven for about 3 to 5 minutes or until the cheese is melted.
Nutrition:
Calories: 562; Fat: 19.9 g; Cholesterol: 60 mg; Carbohydrates:66.7 g; Protein: 31.6 g; Iron: 1 mg; Calcium: 409 mg; Potassium: 370 mg.
Tips
Plain tomatoes may be substituted for cherry tomatoes.

Chicken Alfredo Pita Pizza

Preparation Time: 40 Minutes
Servings:4

Ingredients Needed

- Frozen chicken tenders – 6 smalls (thawed and sliced)
- Garlic humus – ¼ cup
- Basil pesto – 4 teaspoons
- Fresh spinach leaves – 1 cup (chopped)
- Olive oil (divided)
- Garlic salt – 1 pinch or according to taste
- Pita bread rounds – 4
- Prepared Alfredo sauce – ½ cup
- Marinated artichoke hearts – 1 (6.5 ounce) jar (drained and chopped)
- Feta cheese – ¾ cup (crumbled)
- Parmesan cheese – ½ cup (shredded)
- Mozzarella cheese – ¾ cup
- Fresh mushrooms – ½ cup (sliced)

Instructions to follow

Preheat oven to 350 degrees F (175 degrees C)

In a skillet and over medium heat, heat 1 tablespoon of olive oil. Season chicken with garlic salt the cook and stir in hot oil for about five minutes or until the chicken is no longer pink in the middle. Set aside to cool.

Spread 1 tablespoon of hummus over one side of each of the pita rounds so that it is nearly touching the edges. Top with layers pesto and Alfredo sauce.

After this, top with even portions of chicken, artichoke hearts, feta cheese, mozzarella cheese, parmesan cheese and mushrooms.

Drizzle the remaining olive oil on the pizzas and bake in the preheated oven for about 15 minutes or until the cheese is melted and the crust on the pitas is slightly brown.

Nutrition:

Calories: 707; Fat: 38.7 g; Cholesterol: 137 mg; Carbohydrates: 39.8 g; Protein: 50.9 g; Iron: 3 g; Calcium: 521 mg; Magnesium: 72 mg; Potassium: 463 mg.

Tips

You can always experiment with the ingredients by substituting them with other ingredients that are closely related to them.

Apple and Feta Pan Fried Pizzas

Preparation Time: 35 Minutes
Servings: 8)
Ingredients Needed
Hot water – ½ cup
Feta cheese – 8 ounces (crumbled)
Fresh thyme – 1 tablespoon (chopped)
Apples – 4 (cored and chopped)
Dy pizza crust mix – 6 ½ ounces
Olive oil – 5 tablespoons
Red onion – 1 (thinly sliced)
Butter – ½ teaspoon
Ground black pepper to taste.
Instructions to follow
Combine the contents of the pizza dough package with ½ cup of hot water in a medium-sized bowl. Stir vigorously and set the bowl in a warm place of about 85 degrees F or 35 degrees C for 5 minutes.
Turn the dough onto a floured board and divide into 8 small sections, then knead and shape into rounds.
Heat the olive oil in a large skillet, add the dough and fry until it is lightly browned on both sides.
Once this is cooked, place the circles on a cookie sheet. Sprinkle the feta, red onion and thyme on top of the circles.
Bake the pizza for about 10 to 12 minutes or until the feta begins to brown.
While the pizzas bake, heat ½ tablespoon of butter and a few sprigs of thyme in the previously used skillet. Mix the apples into the skillet and cook until the apples are soft and golden. Put the apples on top of the pizzas, season with pepper and serve.
Nutrition:
Calories: 277; Fat: 15.8 g; Cholesterol: 27 mg; Carbohydrates: 28 g; Protein: 6.9 g; Iron: 1 mg; Calcium: 170 mg; Potassium: 117 mg.

Tips
Instead of a pre-prepared pizza crust, you can always make your own pizza dough and do not deep fry it but pan fry.

Eggplant Pizza

Servings: 4
Preparation Time: 40 minutes
Ingredients:
One large eggplant
⅓ Cup Olive oil
Pepper
Salt
2 cups cherry tomatoes – halved
1½ cups Shredded mozzarella cheese
1¼ cups Marinara sauce
½ cup Torn basil leaves
Instructions:
Preheat your oven to 450°F.
Slice the eggplants into slices and arrange them on the baking sheet.
Cover each of the eggplant slices with olive oil.
Roast the eggplant slices until tender.
Spread two tbsp of the marinara sauce on each piece.
Add the cheese and arrange cherry tomato pieces on each.
Cook for about four minutes, removing when the cheese melts.
Nutrition: calories 466; fat 23.5 g; carbohydrates 44.5 g; protein 20 g

Avocado Tomato Pizza

Servings: 5
Preparation Time: 42 minutes
Ingredients:
Pizza Crust:
Two tbsp Olive oil
Two cloves of garlic minced
1 1/4 cup Chickpea flour
Sea salt and pepper
1 1/4 cup Cold water
One tsp any herbs of choice
Soccer Pizza Toppings:
Extra salt/pepper for seasoning
1 Roma tomato sliced
One half of an avocado
2 ounces of gouda (sliced thin)
1/3 cup Tomato sauce
Three tbsp Green onion (chopped)
Red pepper flakes
Instructions:
Preheat your oven at 350 F.
Place pan in oven to warm.
Mix olive oil, water, flour, and herbs until smooth and let it chill for ten minutes.
Remove pan from oven after ten minutes then add a tbsp of oil to the pan, moving it around to cover the bottom of the pan then pour in your batter.
Set the oven temperature to 425F and put the pan back in the oven, until the mixture has set.
Remove pan from oven and spread the tomato sauce on top, then sliced tomato and avocado.
Put the Gouda slices on top of the tomato and avocado.
Put back in oven until the cheese melts.
Let it cool then put the fresh green onion on the top.

Drizzle a bit olive oil on top and enjoy.
Nutrition: calories
212; fat11.3 g; carbohydrates21.5 g; protein6.2 g

Watermelon Feta and Balsamic pizzas

Servings: 4
Preparation Time: 10 minutes
Ingredients:
One watermelon (sliced 1-inch thick)
1-ounce crumbled Feta cheese
Six sliced olives
One tsp Mint leaves
1/2 tbsp Balsamic glaze
Instructions:
Cut the watermelon slice into quarters.
Place them on a round dish and place cheese, olives, balsamic glaze, and mint leaves on top.
Enjoy.
Nutrition: calories 25; fat 1.2 g; carbohydrates 3.1 g; protein 0.9 g

Spinach and Sun-Dried Tomato Pizza

Servings: 4
Preparation Time: 25 minutes.
Ingredients
½ cup dry-packed sun-dried tomatoes
2 tbsp fresh basil
3 tbsp parmesan cheese
⅓ cup tomato juice
One pizza dough base
1 tsp olive oil
Two cloves garlic
1 tbsp balsamic vinegar
2 tbsp tomato paste
2 cups fresh spinach leaves
Instructions
Soak the sun-dried tomatoes in hot water for about ten minutes. Blend the tomatoes, tomato juice, tomato paste, basil, vinegar, olive oil, and garlic and parmesan cheese until smooth.
Spread the sauce on the pizza base and top it with spinach leaves and mozzarella sprinkled on top.
Bake until cheese melts.
Nutrition: Calories: 281 calories; fat 19.7 g; carbohydrates 43.9 g; protein; 79 14.8 g

Mediterranean Pizza Omelet
Serves 4
Preparation Time: 15 minutes
Ingredients
8Eggs
One tsp Oregano
One tbsp extra-virgin olive oil
Four tbsp Tomato Passata sauce
Ten pitted and sliced Black olives

2 cups shredded Cheddar cheese
Pinch of Salt
Pinch of Pepper
Instructions
In a bowl whisk together eggs, oregano, and a pinch of salt and pepper.
Heat olive oil in a skillet on medium-low heat and pour eggs tipping the pan around until the eggs set.
When the eggs set, pour the Passata sauce over then add the sliced olives over the sauced area.
Sprinkle the cheddar cheese on top.
Place a lid on the skillet and serve when ready.
Nutrition: calories 375, fat 9 g, carbohydrates 58 g, protein 17 g,

Broccoli Pizza

Serving: 6
Preparation Time: 30 minutes
Ingredients:
1 pound of pizza dough
5 ounces arugula
1/2 cup of pesto
Salt
Ground pepper
2 cups broccoli
1/4 cup of water
1 cup of mozzarella cheese
Instructions
Preheat oven to 400°F
Transfer the dough to the baking sheet then bake until it is crisp on the bottom.
Cook the broccoli and water in a pan until the broccoli is tender.
Mix in arugula and cook, until withered.
Add seasoning and spread pesto evenly on the crust, then, top with the broccoli mixture and cheddar.
Prepare until the cheddar is liquefied and serve.

Nutrition: 323 calories, fat 13 g, carbohydrates 33 g, protein, 15 g,

Chicken and Cheese Pizza

Serving: 4
Preparation Time: 50 minutes.
Ingredients:
Two boneless chicken breasts
1/2 cup of salad dressing
One can of pizza crust
1 cup of tomatoes
1/4 cup of onions
1 cup of mozzarella cheese
1 cup of cheddar cheese
Instructions:
Preheat the oven to 220 degrees C.
Put chicken in a lined pan and cook until the juices run clear. Cool, then shred into little pieces.
Press the dough into the arranged pizza pan and bake until it starts to tan.
Remove from oven and spread salad dressing over the crust. Sprinkle on mozzarella and cheddar cheese and top with tomatoes, onion, and chicken.
Sprinkle more of cheddar cheese and return to oven until cheddar is liquefied.
Serve and enjoy.
Nutrition: calories 326, fat 16 g, carbohydrates 30 g, protein 18 g,

Mushroom and Ricotta Pizza

Serving: 3
Preparation Time: 1hour 10 minutes.
Ingredients:

Olive oil
One onion
Six mushrooms
Kosher salt
Sugar for taste
One tsp vinegar
1 cup shredded Asiago cheese
2/3 cup of ricotta cheese
Two tortillas
Grounded black pepper
Instructions:
Preheat the oven to 450°F.
Heat 1 tbsp olive oil in a dish and add the cut onions, a squeeze of sugar and the vinegar cooking for three minutes until the onions are translucent.
Place a tortilla on a lined baking sheet and brush each with olive oil.
Sprinkle every tortilla with a large portion of a shredded Asiago cheese and bits of ricotta cheese.
Sprinkle with mushrooms, salt, pepper, and caramelized onions and place the baking sheets in the oven.
Heat until the crust is tan.
Serve and enjoy.
Nutrition: calories 329, fat 10 g, 30 g of carbohydrates, protein 18 g

Tomato and Pepper Pizza

Serving: 4
Preparation Time: 40 minutes.
Ingredients:
One medium zucchini
Two tsp fresh oregano
1 pint of small tomatoes
Two tbsp of tomato puree
Eight basil leaves

Three tbsp of parmesan cheese
1-pound whole wheat pizza dough
4 ounces mozzarella cheese
1/4 tsp of salt
1/2 tsp of grounded pepper
Cornmeal
One yellow bell pepper
Instructions:
Preheat oven to 400F.
Cook zucchini for around 4 minutes till ready.
In Pa food processor, pulse tomatoes, tomato puree, basil, oregano, salt, and pepper until it becomes smooth.
Sprinkle cornmeal onto a baking sheet.
Transfer the dough to a baking sheet, with the underside covered with cornmeal.
Bake for 7 minutes.
Spread the tomato mixture and layer with mozzarella, pepper, and the zucchini and finally sprinkle on the parmesan.
Bake until the cheddar has softened, and the base of the crust has tanned.
Enjoy.
Nutrition: calories 375, fat 9 g, carbohydrates 58 g, protein 17 g,

Tomatoes and Pesto Pizza
Serving Size: 5
Preparation Time: 30 minutes.
Ingredients:
1 pound of pizza dough
Grounded pepper
1/2 cup of prepared pesto
Four ripe tomatoes
Salt for taste
1/4 cup of fresh basil leaves
1/2 cup of feta cheese
Instructions:
Preheat the oven.

Place the dough mixture on a floured surface and create four crusts.
Lay the crusts and topping and garnishes and then heat in the oven.
Bake until the crust is delicately brown.
Spread pesto and top with tomatoes and later with feta and pepper and bake again until ready.
Sprinkle with basil and serve while still hot.
Nutrition: calories 430, fat 16 g, carbohydrates 60 g, protein, 13 g.

Pasta Salad
Preparation Time: 25 Minutes
Servings: 2)
Ingredients
Macaroni – 1 cup
Black olives – ¼ cup (sliced)
Olive oil – 1 tablespoon
Lemon juice – 1 teaspoon
Red bell peppers – 2 cups (diced and roasted)
Feta cheese – ¼ cup (crumbled)
Garlic – 1 tablespoon (minced)
Salt and pepper to taste.
Instructions:
Combine olive oil and chopped garlic in a small bowl in a small bowl or cup before setting aside.
In a large pot, cook pasta until it is soft and firm, then drain.
Lastly, transfer the pasta to a medium mixing bowl, add, olives, roasted red peppers and feta cheese. Toss together with the olive oil mixture and lemon juice. Season with salt and pepper. Then serve immediately.
Nutritional facts:
Calories: 340; Fat: 13.6 g; Cholesterol: 17 mg; Sodium: 472 mg; Carbohydrates: 44.2 g; Protein: 10.3 g; Iron: 3 mg; Calcium: 130 mg; Potassium: 150 mg.
Tips

If you are trying this recipe for the first time, it is advisable that you use less garlic.

Pasta with Greens

Preparation Time: 35 Minutes
Servings: 8

Ingredients
Swiss chard – 1 bunch (remove the stems)
Oil packed sun-dried tomatoes – ½ cup (chopped)
Green olives – ½ cup (chopped and pitted)
Fresh parmesan cheese – ¼ cup (grated)
Dry fusilli pasta – 1 (16 ounce) package
Olive oil – 2 tablespoons
Kalamata olives – ½ cup (chopped and pitted)
Garlic – 1 clove (minced)
Instructions
Cook pasta in lightly salted water for 10 to 12 minutes until al dente then drain.
Put the chard in a microwave safe bowl, fill with water until it is about ½ filled with water. Cook on high in the microwave for about 5 minutes until the chard is limp then drain.
Over medium heat, heat the oil in a skillet. Stir in the oil, the sun-dried tomatoes, green olives, kalamata olives and garlic. Mix in the chard the cook and stir until the mixture is tender. Toss with the pasta and sprinkle with parmesan cheese to serve.
Nutritional fact
Calories: 296; Fat: 9.7 g; Cholesterol: 2 mg; Sodium: 467 mg; Carbohydrates: 44.6 g; Protein: 9.6 g; Calcium: 66 mg; Iron: 3 mg; Potassium: 329 mg.
Tips
You can substitute the pasta with another any other that you like.

Harvest Pasta

Preparation Time: 35 Minutes
Servings: 6
Ingredients
Kalamata olives – 1/3 cup (pitted)
Garlic – 2 cloves (minced)
White sugar – 1 tablespoon or more to taste
Dried oregano – 1 teaspoon
Vegetarian burger crumbs – ¾ cup
Diced tomatoes – 2 (14.5 ounce) cans
Bottled roasted red peppers – 1/3 cup (chopped)
Balsamic vinegar – 1 ½ tablespoons
Olive oil – 2 tablespoons
Black pepper to taste
Penne pasta – 1 pound
Instructions:
In a large saucepan, stir the olives, garlic, sugar, oregano, tomatoes, red pepper, vinegar. Bring this to simmer for about 20 to 30 minutes over medium high-heat before reducing to medium-low and let simmer until the sauce starts to thicken.
In a large pot, pour lightly salted water and boil over high heat. Once the water is boiling, put in the penne pasta and leave to boil.
Cook the pasta uncovered for about 11 minutes and remember to stir occasionally until the pasta is al-dente. After this drain.
Once the tomato sauce is done, pour it into the blender no more than halfway full. Hold down the lid and carefully start the blender using a few pulses to get the sauce moving before leaving it on to puree. Afterwards, puree until the mixture is smooth, then return to the pot.
Stir in the burger crumbs and simmer until it is hot. Then pour the finished sauce over the penne pasta to serve.
Nutritional facts:
Calories: 392; Fat: 8.8 g; Cholesterol: 0 mg; Carbohydrates: 64.9 g; Protein: 13.4 g; Iron: 6 mg; Calcium: 72 mg; Potassium: 345 mg.
Tips

You can also use a stick blender to puree the sauce in the pot until it is smooth.

Pollo Mediterranean

Preparation Time: 35 Minutes
Servings: 4
Ingredients
Olive oil – 2 tablespoons
Garlic – 3 cloves (minced)
Ground black pepper – ½ teaspoon
Sun-dried tomatoes packed in oi – ¼ cup (chopped and drained)
Dry white wine – ½ cup
Chicken tenders – 12 (sliced into strips)
Salt – ½ teaspoon
Italian seasoning – 1 tablespoon
Green olives – 2 tablespoons (sliced)
Fresh parsley – 2 tablespoons (chopped)
Sour cream – ½ cup
Salt – ½ teaspoon
Milk – 1 cup
Cornstarch – 1 ½ teaspoons
Water – ¼ cup
Instructions
In a skillet and over medium heat, heat olive oil. Place chicken and garlic in the pan. Season with pepper, Italian seasoning and ½ teaspoon of salt.
Stir in the olives, wine, parsley, tomatoes and olives then reduce heat to a low and continue cooking until the chicken is no longer pink at the center. Remove and place chicken on a late with the sauce still in the pan. Stir into the remaining sauce ½ teaspoon of sauce.
In a small bowl, whisk cornstarch and water together. Increase heat to the medium and whisk in the cornstarch mixture.

Continue stirring until the sauce has thickened. Serve the sauce with chicken.
Nutritional fact:
Calories: 392; Fat: 19.7 g; Cholesterol: 111 mg; Carbohydrates: 9.2 g; Protein: 38 g; Calcium: 157 mg; Potassium: 590 mg.
Tips
You can use artichoke in the cooking.

Pasta Fagioli Soup

Preparation Time: 75 Minutes
Servings: 8
Ingredients:
Water – 3 cups
Crisp cooked bacon – 8 slices (crumbled)
Dried parsley- 1 tablespoon
Garlic – 1 tablespoon (minced)
Garlic powder – 1 teaspoon
Ground black pepper – ½ teaspoon
Salt- 1 ½ teaspoon
Dried basil – ½ teaspoon
Tomato sauce – 1 (8 ounce) can
Seashell pasta – ½ pound
Great Northern beans – 2 (14 ounce) cans (undrained)
Chicken broth – 2 (14.5 ounce) can
Diced tomatoes – 1 (29 ounce) can
Chopped spinach – 1(14 ounce) can (drained)
Instructions
Combine all the other ingredients apart from pasta in a large stock pot to cook and boil. Let simmer for about 40 minutes. Add pasta and cook with the pot uncovered until the pasta is tender. This should take approximately 10 minutes.
Serve.
Nutrition

Calories: 288; Fat: 3.6 g; Cholesterol: 7 mg; carbohydrates: 48.5 g; Protein: 15.8 mg; Iron: 5 mg; Calcium: 100 mg; Potassium: 701 mg

Tip
You can substitute half of the canned ingredients for better nutritional outcomes.

Pasta al Mediterraneo

Preparation Time: 27 Minutes
Servings: 6
Ingredients
Perciatelli pasta – 1 pound
Pine nuts – 3 tablespoons (lightly roasted)
Fresh parsley – 2 tablespoons (chopped)
Lemon – 1 (juiced)
Can tuna – 2 (5 ounce) package (drained)
Kalamata olives – 12 (pitted and sliced)
Garlic – 1 clove (crushed)
Fresh basil – 4 ounces (chopped)
Olive oil – 6 tablespoons
Feta cheese – 2 ounces (optional)
Instructions
Cook pasta in a large bowl of slightly salted water until al dente. Meanwhile, mix in a large bowl, olives, garlic, basil, tuna, pine nuts, parsley and crumbled feta cheese.
Drain the pasta. If the plan is to serve cold, then rinse the pasta with cold water until it is no longer hot. In a large bowl, place pasta together with lemon juice and olive oil. Stir into the pasta mixture, the tuna mixture.
Serve hot or cold.
Nutritional fact
Calories: 519; Fat: 22 g; Cholesterol: 21 mg; Sodium: 255 mg; Carbohydrates: 59.5 g; Protein: 24.2 g; Calcium: 122 mg; Potassium: 370 mg.
Tips.

If possible, use fresh lemon juice instead of bottled ones.

Tomato Basil Penne Pasta

Preparation Time: 45 Minutes
Servings: 4
Ingredients
Basil oil – 1 tablespoon
Garlic – 3 cloves (minced)
Pepper jack cheese – 1 cup
Parmesan cheese – ¼ cup (grated)
Basil oil – 1 tablespoon
Grape tomatoes – 1 pint (halved)
Mozzarella cheese – 1cup (shredded)
Fresh basil – 1 tablespoon (minced)
Instructions
Over high heat, bring a large pot of water to boil. Cook pasta in the boiling water for about 11 minutes until al dente, then drain. In a large skillet and over medium-high heat, heat the basil and olive oil. Cook garlic in oil until soft. Afterwards, add tomatoes, reduce the heat to a medium and leave to dimmer for 10 minutes.
Stir in the mozzarella, parmesan cheese and pepper jack. When the cheese begins to melt, mix in the cooked penne pasta. Season with fresh basil.
Nutritional fact
Calories: 502; Fat: 24.8 g; Cholesterol: 58 mg; Sodium: 462 mg; Carbohydrates: 47.1 g; Protein: 24.1 g; Calcium: 474 mg; Potassium: 311 mg.
Tip
If basil oil is unavailable, use 2 tablespoons of olive oil.

Whole Wheat Pasta Toss

Preparation Time: 45 Minutes
Servings: 8
Ingredients
Olive oil – 1/3 cup
Marinated artichoke hearts – 1 (8 ounce) jar (drained)
Kalamata olives – ¼ cup (pitted and quartered)
Feta cheese – ½ cup (crumbled)
Whole wheat penne pasta – 1 (1 pound) package
Garlic – 4 large cloves (pressed)
Pickled red peppers – 7 (cut into strips)
Fresh spinach leaves – 2 cups
Instructions
Fill a large bowl with lightly salted water and bring to boil. Put in the penne and continue to boil. Cook the pasta uncovered, stirring occasionally for 8 minutes or until al dente, then drain. In a large non-stick skillet and over medium heat, heat olive oil, the cook and stir in garlic into the hot oil for about 30 seconds until it is fragrant, for about 5 minutes. Gently fold the spinach into the mixture and stir just until slightly wilted and dark green.
Remove the mixture from heat and stir in the penne pasta until it is thoroughly combined; lightly toss pasta mixture in with the feta steam, cover the skillet with a lid and let the vegetables and pasta steam for about 10 minutes before serving.
Nutritional fact
Calories: 367; Fat :14.7 g; Cholesterol: 8 mg; Sodium: 347 mg; Carbohydrates: 47.4 g; Protein: 12.9 g; Iron: 1 mg; Calcium: 60 mg; Potassium: 58 mg.

Quick Mediterranean Pasta

Preparation Time: 25 Minutes
Servings: 6
Ingredients

Breadcrumbs – ¼ cup
Dried basil – 1 teaspoon
Spaghetti – 8 ounces
Dried oregano – 1 teaspoon
Olive oil – 1 tablespoon
Instructions
Boil slightly salted water in a large pot, put spaghetti in it and cook until al dente. Rinse and cool with water, then drain well. Mix the breadcrumbs, basil, oregano and cooked pasta in a large bowl. Pour as much olive oil as you would like over the mixture and serve.
Nutritional facts
Calories: 178; Fat: 3.1 g; Cholesterol: 0 mg; Sodium: 35 mg; Carbohydrate: 31.4 g; Protein: 5.5 g; Iron: 2 mg; Calcium: 25 mg; Potassium: 104 mg.
Tips
You can always experiment with the recipe

Mediterranean Fish and Pasta Stew

Preparation Time: 50 Minutes
Servings: 6
Ingredients
Onions – 2 (chopped)
Crushed tomatoes – 1 (28 ounce) can
Fresh parsley – ½ cup (chopped)
Worcestershire sauce – 2 tablespoons
Paprika – 1 teaspoon
Dry pasta – 3 ounces
Garlic – 4 cloves (minced)
Olive oil – 1 tablespoon
Water – 6 cups
Fresh cilantro – ½ cup (chopped)
Ground cinnamon – 1 teaspoon
Cod fillets – 1 ½ pounds (cubed)
Salt to taste

Ground black pepper – 1 tablespoon

Instructions

In a large pot, sauté the onions and garlic in the olive oil for 5 minutes over medium heat while stirring constantly. Add tomatoes with the liquid, parsley, water and cilantro. Bring the mixture to boil and reduce heat to low and simmer for about 15 minutes.

Stir in the Worcestershire sauce, paprika, cinnamon and fish, the simmer over medium heat for 10 minutes. Add the pasta and simmer for about 8 minutes more or until the pasta is tender. Season with salt and ground pepper to taste.

Nutritional facts

Calories: 237; Fat: 4.2 g; Cholesterol: 66 mg; Sodium: 300 mg; Carbohydrates: 26.2 g; Protein: 25.3 g; Iron: 4 mg; Calcium: 103 mg; Potassium: 1030 mg.

Tips

You can substitute some ingredients and add in some more in accordance with your taste.

Parsley Pesto Paste

Servings: 4
Preparation Time: 5 minutes

Ingredients

2 cups of parsley leaves
1/2 cup of grated parmesan cheese
Two cloves of garlic
1/2 cup lemon juice
1/4 cup olive oil
1/3 cup pine nut
Table salt to taste

Instructions

Put all ingredients except the parmesan cheese in a food processor then pulse until smooth.
Remove from the blender, add grated parmesan and gently stir. Serve.

Nutrition: Calories 266, Fat 25g, Carbohydrates 6g, Protein 8g

Potato in Tomato Paste

Servings: 4
Preparation Time: 40 minutes
Ingredients
Four large cubed potatoes
1 Tbsp aromatic dry spices mix
One onion, chopped
4 Tbsp Olive oil
Black pepper
One minced garlic clove
1 cup tomato paste
1 cup of water
Chopped parsley,
Salt
Instructions
Heat the olive oil in a pan over medium heat and sauté the onion until translucent.
Add the potatoes, the spice mixture and continue to sauté.
Add the garlic, tomato paste, diced tomato, water, salt and pepper, and stir.
Cover the pot and cook for half an hour over low heat.
Serve with fresh coriander.
Nutrition Facts: Calories 312, Fat 14g, Carbohydrates 43g, Protein 6g

Hummus
Servings: 14
Total time 20 minutes
Ingredients
1/2 cup tahini
1 tsp salt
Two cloves garlic halved

1 tbsp olive oil
2 cup canned garbanzo beans, drained
1/2 cup lemon juice
1 tbsp paprika
1 tsp parsley
Instructions
Pulse the garlic, lemon juice, garbanzos, salt, and tahini in a food processor until smooth.
Add this to a bowl with olive oil, paprika, and parsley.
Enjoy.
Nutrition: Calories 77 Fat 4.3 g Carbohydrates 8.1g Protein 2.6 g

Hollandaise Sauce

Servings: 1
Preparation Time: 10 minutes
Ingredients
One lemon (Zested and juiced)
One tsp garlic powder
1/2 tsp cayenne pepper
1/2 cup cashew butter
Two tsp Dijon mustard
1/2 cup of warm water
1/2 tsp ground turmeric
Instructions
In a food processor, put all ingredients, and then pulse until smooth.
Put it in a sealed container and refrigerate it for up to three days.
Enjoy.
Nutrition: Calories 150 Protein 6 g Fat 12 g Carbohydrates 10 g

Creamy Tahini Dip

Servings: 4

Preparation Time: 5 minutes
Ingredients
Half a lemon (Juiced)
One crushed garlic clove
Salt
1/2 cup tahini
2 cups of water
Fresh parsley, chopped
Black pepper
Instructions
Put the tahini, salt, lemon juice, garlic, and a little water in a bowl then stir until the tahini becomes white and smooth. Sprinkle the parsley and black pepper and serve.
Enjoy.
Nutrition: Calories 93 Protein 2.6 g Fat 8.1 g Carbohydrates 4.4 g

Basil Lime Dip

Servings: 16
Preparation Time: 10minutes
Ingredients
Ten garlic cloves, crushed
1/4 cup brown rice syrup
8 ounces hemp oil
One tsp of sea salt
One pinch xanthan gum
1 1/2 cups chopped basil,
Six tbsp key lime juice
Instructions
In an airtight jar, put all the ingredients except the xanthan gum, and then shake to well.
Put the mixture plus the xanthan, into a blender and pulse.
Return the mixture in the jar.
Enjoy.
Nutrition: Calories 143 Cholesterol 0 mg Fat 14 g Carbohydrates 6 g

Cilantro Dip

Servings: 7
Preparation Time: 4minutes
Ingredients:
12 cloves of garlic
4 cups cilantro leaves
One tsp salt
1/2 tsp ground black pepper
1 cup olive oil
Instructions:
Add all ingredients to a blender and pulse until velvety.

You can put in the refrigerator for up to two days.
Enjoy.
Nutrition: calories 230; fat 20.5 g; carbohydrates 7.1 g; protein 5 g

Tahini Sauce

Servings: 6
Preparation Time: 7 minutes
Ingredients:
Four mashed garlic cloves
Salt to taste
1 cup tahini paste
1/2 cup lemon juice
Seven tbsp water
Instructions
Put all ingredients in a bowl and whisk until well combined.
Refrigerate up to 5 days.
Enjoy.
Nutrition: calories 77; fat 6.6 g; carbohydrates 3.2 g; protein 2.3 g

Arugula Salsa

Servings: 7
Preparation Time: 30 minutes
Ingredients:
30 Kalamata olives, pitted, quartered
Three tbsp olive oil
One chopped red bell pepper
One chopped yellow bell pepper
Two tsp fennel seeds, crushed
1 cup baby arugula, chopped
Instructions
Heat oil in a pan over medium heat.
Add fennel seeds and sauté until fragrant.
Add bell peppers and sauté until they are soft.
Transfer into a bowl.
Add salt, pepper, and arugula and stir until arugula wilts.
Enjoy.
Nutrition: calories 16; fat 0.1 g; carbohydrates 3.9 g; protein 0.6 g

Exotic Dip

Servings: 10
Preparation Time: 25 minutes
Ingredients
Three tbsp melted coconut oil
1 cup fresh baby spinach leaves
1 tsp salt
1 3/4 cups sour cream
1/2 cup softened cream cheese
One tbsp chopped dill
One tsp grated lemon peel

1/3 cup of toasted pine nuts,
1 tsp minced garlic
½ cup crumbled feta cheese
Two medium green onions, chopped
Instructions
Place a pot over medium flame and heat butter until it melts.
Add garlic and cook until fragrant.
Stir in spinach and salt then cook for five minutes until the leaves wither.
Drain the spinach into a colander and squeeze until dry thoroughly.
In a blender, place spinach mixture, lemon peel, sill, sour cream, cream cheese, feta cheese, and pulse until smooth.
Transfer into a bowl and place it into the refrigerator for 2 hours.
Garnish with chopped onion, pine nuts, and lemon peel.
Enjoy.
Nutrition: calories 142; fat 12.9 g; carbohydrates 2.4 g; protein 4.5 g

Conclusion

Thank you for choosing and reading this book. We hope it was useful and able to provide you with a thorough overview of the Mediterranean diet, its lifestyle, its dos and don'ts. We hope this book has helped you understand more about this diet and given you the tools to set off on further research, to learn about the background of this diet and the food pyramid it is based on. Please remember, however, not to undergo any significant lifestyle or dietary changes without consulting your GP as there may be contraindications to certain elements. If this isn't your case, the Mediterranean diet will surprise you with its beneficial effects on your health, including but not limited to improving the appearance of your skin, lowering your cholesterol levels, helping prevent the onset of type 2 diabetes... all this in addition to losing weight!

© **Copyright 2020 - All rights reserved.**

The content contained within this book may not be reproduced, duplicated or transmitted without direct written permission from the author or the publisher.

Under no circumstances will any blame or legal responsibility be held against the publisher, or author, for any damages, reparation, or monetary loss due to the information contained within this book. Either directly or indirectly.

Legal Notice:

This book is copyright protected. This book is only for personal use. You cannot amend, distribute, sell, use, quote or paraphrase any part, or the content within this book, without the consent of the author or publisher.

Disclaimer Notice:

Please note the information contained within this document is for educational and entertainment purposes only. All effort has been executed to present accurate, up to date, and reliable, complete information. No warranties of any kind are declared or implied. Readers acknowledge that the author is not engaging in the rendering of legal, financial, medical or professional advice.

The content within this book has been derived from various sources. Please consult a licensed professional before attempting any techniques outlined in this book.

By reading this document, the reader agrees that under no circumstances is the author responsible for any losses, direct or indirect, which are incurred as a result of the use of information contained within this document, including, but not limited to, — errors, omissions, or inaccuracies

www.ingramcontent.com/pod-product-compliance
Lightning Source LLC
Chambersburg PA
CBHW071357210526
45465CB00001B/132